INTEGRALS AND SUMS

Some New Formulae
for their Numerical Evaluation

INTEGRALS AND SUMS

*Some New Formulae
for their Numerical Evaluation*

by

P. C. CHAKRAVARTI

UNIVERSITY OF LONDON
THE ATHLONE PRESS
1970

Published by
THE ATHLONE PRESS
UNIVERSITY OF LONDON
2 Gower Street London WC1
Distributed by Tiptree Book Services Ltd
Tiptree, Essex

Australia and New Zealand
Melbourne University Press

USA
Oxford University Press Inc.
New York

0 485 11114 4

Printed in Great Britain by
JOHN WRIGHT AND SONS LTD.
AT THE STONEBRIDGE PRESS, BATH ROAD, BRISTOL BS4 5NU

To My Parents

PREFACE

The Euler–Maclaurin formula

$$h \sum_{r=0}^{n-1} f(x_r + sh) - \int_{x_0}^{x_n} f(t)\, \mathrm{d}t = \sum_{r=0}^{\infty} h^{r+1} \frac{B_{r+1}(s)}{(r+1)!} \{f^{(r)}(x_n) - f^{(r)}(x_0)\}$$

originated during the years 1730–1740 from the independent discoveries of L. Euler and C. Maclaurin, since when it has attracted the attention of many mathematicians. It relates a sum and an integral and gives explicitly the correction terms which allow one to be converted into the other. Consequently, it has been used for summation as well as for approximate integration. Further, it has in fact provided some of the more remarkable results in the theory of divergent series and asymptotic expansions.

It was not until 1823 that the Euler–Maclaurin formula was placed on a proper footing by S. D. Poisson, who obtained, by the aid of Fourier's theorem, an explicit expression for the remainder term. Other formulations were made, for example, by Schloemilch in terms of the Bernoulli functions (i.e. the periodic continuations of the Bernoulli polynomials), by Glaisher using the method of integration by parts, by Kronecker using contour integrals and the theory of residues, and by Boole using his operational method.

The Newton–Cotes and Gauss formulae have been extensively generalized, especially with respect to the inclusion of weighting functions. The methods for constructing such formulae and the general issues involved are well formulated and clearly understood. Similar overall developments of the Euler–Maclaurin formula, however, have so far been lacking.

This volume presents the results of the author's researches, carried out over a period of years, to fill this gap.

University College London

P. C. C.

ACKNOWLEDGEMENTS

The author would like to express his gratitude to Professor W. G. Bickley without whose continuous encouragement and guidance this work would not have been possible.

He is also indebted to Dr B. Spain for his most generous help, to Dr J. C. P. Miller for his interest and helpful suggestions, and to the Sir John Cass Foundation for granting a Research Assistantship for the period October 1958 to September 1961, when part of this work was done.

Sincere thanks are due to Dr M. Bernal for discussions and criticisms, and to Dr G. deBarra for checking the final draft. A special note of appreciation is offered to the staff of The Athlone Press for their invaluable editorial help.

CONTENTS

IV. ARBITRARY CORRECTION POINT FORMULAE AND FORMULAE WITH CORRECTION TERMS EXPRESSED IN TERMS OF QUANTITIES OTHER THAN DERIVATIVES

CHAPTER I

COMPLEX VARIABLE METHOD

Introduction

The aim of the present work is twofold: to derive three different types of formulae for the numerical evaluation of integrals and sums, as well as to develop, in each case, alternative methods for their derivation.

The work is divided into four chapters.

The present chapter is devoted to the use of complex variable methods for the derivation of general rectangular formulae with prescribed weighting functions.

For this purpose we first of all derive the *complex error lemma* which expresses the 'error' in approximating the integral of a function by its general n-ordinate rectangular sum in terms of three integrals in the complex plane (Section 1). The lemma is believed to be unknown in its general rectangular form (1.1), although the trapezoidal form (1.7) is well known (Hardy, 1956, p. 339).

With the help of this lemma we then proceed to prove that, for

$$F(x) = w_0(x)f_0(x) = w_n(x)f_n(x),$$

where the weighting functions $w_0(x)$ and $w_n(x)$, associated respectively with the two end points x_0 and x_n of the interval of integration, are suitably prescribed, there *exist* formulae of the form

$$h \sum_{r=0}^{n-1} F(x_r + sh) - \int_{x_0}^{x_n} F(x)\, \mathrm{d}x = \sum_{r=0}^{m_n} \frac{\phi_r^n(h, s)}{r!} f_n^{(r)}(x_n) - \sum_{r=0}^{m_0} \frac{\phi_r^0(h, s)}{r!} f_0^{(r)}(x_0) + R,$$

which connect the integral and the general n-ordinate rectangular sum of a function $F(x)$ with the help of two associated sums of 'correction terms' and a corresponding error term R. The correction terms, we may notice, involve the values of the quotient functions $f_0(x) = F(x)/w_0(x)$ and $f_n(x) = F(x)/w_n(x)$, and their derivatives, at the corresponding end points of the interval of integration.

The analysis also provides us with (1) an analytical expression for the error term R as well as (2) an explicit formula for the calculation of the correction coefficients $\phi_r^0(h, s)$ and $\phi_r^n(h, s)$ (Section 2).

In the remaining sections the formulae of Section 2 are applied to obtain in a convenient form the special formulae for a number of important weighting functions.

The trapezoidal formula—which, strictly speaking, is not a special case of the rectangular formula—is treated throughout as a corollary to the general case.

In Sections 3 and 4 the form of the correction coefficients for the special weighting functions e^{px}, $\sin(\beta x)$, $\cos(\beta x)$, $\sinh(\alpha x)$ and $\cosh(\alpha x)$ are established with the help of simple analytical techniques; a different treatment based on the analytical properties of the higher transcendental Φ-function is given in Appendix A. The sections dealing with the relevant error bounds, the comparison with other quadrature formulae and numerical examples are deferred to Chapter II where the formula (3.7) with the exponential weighting function is re-derived with a different expression for the remainder term. The properties of the associated correction coefficients and the formulae for their computation are also developed later in Sections 3–6 of Chapter II, which are devoted to the closely related D-functions (see (3.10)).

In the remaining sections of the present chapter attention is given to the special weighting functions $\sin(\beta x)/x$ and x^{α}, and a brief description of the method for expressing the correction coefficients for the weighting function $w_0(x) = (x - x_0)^{\alpha_0} e^{px}$ in terms of the higher transcendental function $\Phi(z, \mu, \nu)$ is given in Appendix A. The formula (7.8) for x^{α}, it may be noted, has been obtained in Navot (1961) by a different method and with a different form for the remainder term.

The formulae corresponding to the trapezoidal sum and the mid-ordinate sum are, in all cases, specially displayed. Short tables of the correction coefficients corresponding to the trapezoidal sum formulae are given later in Appendix C.

1. *The complex error lemma*

The *complex error lemma* is presented in its simplest form which it adopts when h is unity; this involves no real loss of generality, since all cases can be reduced to this case by a simple change in the independent variable.

Lemma

If $\phi(\zeta)$, a function of $\zeta = \xi + i\eta$, is analytic everywhere in the rectangular domain $(0 \leqslant \xi \leqslant n, \ -\beta \leqslant \eta \leqslant \beta)$, then, for $0 < s < 1$ and n a positive integer,

$$\sum_{r=0}^{n-1} \phi(s+r) - \int_0^n \phi(\xi)\,\mathrm{d}\xi = i_n(s) - i_0(s) + j(s), \tag{1.1}$$

where $j(s)$ and $i_k(s)$ $(k = 0, n)$ are defined by the equations

$$i_k(s) = \frac{i}{2\pi} \int_0^{2\pi\beta} \left\{ \pi_s(t)\, \phi\!\left(k + \frac{it}{2\pi}\right) - \pi_{-s}(t)\, \phi\!\left(k - \frac{it}{2\pi}\right) \right\} dt, \qquad (1.1a)$$

where

$$\pi_s(t) = \frac{1}{1 - e^{2\pi i s}\, e^t} \qquad (1.1b)$$

and

$$j(s) = \int_0^n \left\{ \frac{\phi(t + i\beta)}{e^{2\pi\beta}\, e^{-2\pi i(t-s)} - 1} + \frac{\phi(t - i\beta)}{e^{2\pi\beta}\, e^{2\pi i(t-s)} - 1} \right\} dt. \qquad (1.1c)$$

For, let

$$\Phi(\zeta) \equiv \frac{\psi(\zeta)}{1 + (-)^{n+1} e^{-2\pi i(\zeta - s)}} + \frac{\psi(-\zeta)}{1 + (-)^{n+1} e^{-2\pi i(\zeta + s)}} \qquad (1.2)$$

and let $\psi(\zeta)$ be analytic everywhere in the rectangular domain

$$(-n/2 \leqslant \xi \leqslant n/2,\ -\beta \leqslant \eta \leqslant \beta).$$

Consider

$$\int_\Gamma \Phi(\zeta)\, d\zeta$$

taken round the closed contour Γ consisting of the rectangle with vertices at $(\pm n/2,\ \pm n/2 + i\beta)$, appropriately indented, with half-circles of radius δ, to exclude all those poles of the integrand which would otherwise lie on Γ. Let Γ' represent the contour Γ minus the arcs of indentation.

Then, by Cauchy's residue theorem,

$$\lim_{\delta \to 0} \int_{\Gamma'} \Phi(\zeta)\, d\zeta = \sum_{r=0}^{n-1} \psi\!\left(-\frac{n}{2} + s + r\right). \qquad (1.3)$$

We also have, if P denotes the Cauchy principal value,

$$P \int_{-n/2}^{+n/2} \Phi(\xi)\, d\xi - \int_{-n/2}^{+n/2} \psi(\xi)\, d\xi = P \int_{-n/2}^{+n/2} [\Phi(\xi) - \tfrac{1}{2}\{\psi(\xi) + \psi(-\xi)\}]\, d\xi = 0, \qquad (1.4)$$

since $\Phi(\xi) - \tfrac{1}{2}\{\psi(\xi) + \psi(-\xi)\}$ is an odd function of ξ.

Therefore, if we decompose Γ' into its four linear segments, we get from (1.3), by virtue of (1.4),

$$\sum_{r=0}^{n-1} \psi\!\left(-\frac{n}{2} + s + r\right) - \int_{-n/2}^{+n/2} \psi(\xi)\, d\xi = \left\{ \int_{n/2}^{n/2 + i\beta} + \int_{n/2 + i\beta}^{-n/2 + i\beta} + \int_{-n/2 + i\beta}^{-n/2} \right\} \Phi(\zeta)\, d\zeta. \qquad (1.5)$$

If we now introduce appropriate changes of variables in the three integrals on the right, then combine together the first and the third integrals along the vertical segments, we obtain

$$\sum_{r=0}^{n-1}\psi\left(-\frac{n}{2}+s+r\right)-\int_{-n/2}^{n/2}\psi(\xi)\,\mathrm{d}\xi$$

$$=\frac{i}{2\pi}\int_{0}^{2\pi\beta}\left\{\Phi\left(\frac{n}{2}+\frac{it}{2\pi}\right)-\Phi\left(-\frac{n}{2}+\frac{it}{2\pi}\right)\right\}\mathrm{d}t-\int_{-n/2}^{n/2}\Phi(t+i\beta)\,\mathrm{d}t,\quad (1.6)$$

which, on writing $\psi(\zeta)=\phi(n/2+\zeta)$, yields the lemma (1.1).

Corollary

For $s=0$ or 1, the lemma (1.1) may be shown to assume the form

$$\tfrac{1}{2}\phi(0)+\sum_{r=1}^{n-1}\phi(r)+\tfrac{1}{2}\phi(n)-\int_{0}^{n}\phi(\xi)\,\mathrm{d}\xi=i_n(0)-i_0(0)+j(0),\quad (1.7)$$

where $i_k(0)$ and $j(0)$ are obtained by writing $s=0$ in (1.1a) and (1.1c) respectively.

2. *The general rectangular formula with prescribed weighting functions*

In order that the formulae to be obtained may be used for the approximate evaluation of an integral of the form

$$\int_{x_0}^{x_n}F(x)\,\mathrm{d}x,$$

where x_0 and x_n are finite, we first introduce the change of variable $z=x_0+h\zeta$ and adopt the notation $x_0+rh=x_r$, so that

$$\int_{x_0}^{x_n}F(x)\,\mathrm{d}x=h\int_{0}^{n}\phi(\xi)\,\mathrm{d}\xi,$$

where $\phi(\zeta)=F(x_0+h\zeta)$.

Also let $F(z)$ be analytic everywhere in the rectangular domain $(x_0\leqslant x\leqslant x_n,\ -b\leqslant y\leqslant b)$.

Then $\beta=b/h$, and the lemma (1.1) assumes the form

$$h\sum_{r=0}^{n-1}F(x_r+sh)-\int_{x_0}^{x_n}F(x)\,\mathrm{d}x=I_n(s)-I_0(s)+J(s)\quad (0<s<1),\quad (2.1)$$

where, for $k=0$ or n,

$$I_k(s)=hi_k(s)=\frac{ih}{2\pi}\int_{0}^{2\pi b/h}\left\{\pi_s(t)\,F\left(x_k+\frac{iht}{2\pi}\right)-\pi_{-s}(t)\,F\left(x_k-\frac{iht}{2\pi}\right)\right\}\mathrm{d}t,\quad (2.2)$$

where $\pi_s(t)$ is as defined before in (1.1b) and

$$J(s)=hj(s)=\int_{x_0}^{x_n}\left\{\frac{F(t+ib)}{\mathrm{e}^{2\pi b/h}\,\mathrm{e}^{-2\pi i(t-x_0-sh)/h}-1}+\frac{F(t-ib)}{\mathrm{e}^{2\pi b/h}\,\mathrm{e}^{2\pi i(t-x_0-sh)/h}-1}\right\}\mathrm{d}t.\quad (2.3)$$

Let us now consider the general case where it is necessary to extract two different weighting functions at the two end points of the range of integration.

Write

$$F(z) = w_0(z)f_0(z) = w_n(z)f_n(z), \tag{2.4}$$

where $w_0(z)$ and $w_n(z)$ are the prescribed weighting functions associated with the two end points of the range, x_0 and x_n respectively.

Then, for $k = 0$ or n,

$$F\left(x_k + \frac{iht}{2\pi}\right) = w_k\left(x_k + \frac{iht}{2\pi}\right)\left\{\sum_{r=0}^{m_k}\left(\frac{iht}{2\pi}\right)^r \frac{f_k^{(r)}(x_k)}{r!} + R_{m_k}\left(x_k + \frac{iht}{2\pi}\right)\right\}. \tag{2.5}$$

Inserting this in (2.2) we obtain

$$I_k(s) = \frac{ih}{2\pi}\int_0^{2\pi b/h}\left[\sum_{r=0}^{m_k}\left\{\pi_s(t)\,w_k\left(x_k + \frac{iht}{2\pi}\right)\right.\right.$$
$$\left.\left. - (-)^r\,\pi_{-s}(t)\,w_k\left(x_k - \frac{iht}{2\pi}\right)\right\}\left(\frac{iht}{2\pi}\right)^r \frac{f_k^{(r)}(x_k)}{r!}\right]dt + \rho_k(s), \tag{2.6}$$

where

$$\rho_k(s) = \frac{ih}{2\pi}\int_0^{2\pi b/h}\left\{\frac{w_k(x_k + iht/2\pi)\,R_{m_k}(x_k + iht/2\pi)}{1 - e^{2\pi is}\,e^t}\right.$$
$$\left. - \frac{w_k(x_k - iht/2\pi)\,R_{m_k}(x_k - iht/2\pi)}{1 - e^{-2\pi is}\,e^t}\right\}dt. \tag{2.7}$$

Equation (2.6) may be written in the form

$$I_k(s) = \sum_{r=0}^{m_k}\left(\frac{ih}{2\pi}\right)^{r+1}\frac{A_{rk}(0, s)}{r!}f_k^{(r)}(x_k) + \sigma_k(s) + \rho_k(s), \tag{2.8}$$

where

$$A_{rk}(\xi, s) = \int_\xi^\infty\left[\frac{(-)^r\,w_k(x_k - iht/2\pi)}{e^{-2\pi is}\,e^t - 1} - \frac{w_k(x_k + iht/2\pi)}{e^{2\pi is}\,e^t - 1}\right]t^r\,dt \quad (0 < s < 1), \tag{2.9}$$

$$\sigma_k(s) = -\sum_{r=0}^{m_k}\left(\frac{ih}{2\pi}\right)^{r+1}\frac{A_{rk}(2\pi b/h, s)}{r!}f_k^{(r)}(x_k), \tag{2.10}$$

and $\rho_k(s)$ is given by (2.7) (provided $w_k(z)$ is such that the integrals $A_{rk}(\xi, s)$ exist).

Using (2.8) in (2.1) we obtain, finally, the *general rectangular formula with prescribed weighting functions* in the desired form

$$h\sum_{r=0}^{n-1}F(x_r + sh) - \int_{x_0}^{x_n}F(x)\,dx$$
$$= \sum_{r=0}^{m_n}\frac{\phi_r^n(h, s)}{r!}f_n^{(r)}(x_n) - \sum_{r=0}^{m_0}\frac{\phi_r^0(h, s)}{r!}f_0^{(r)}(x_0) + R \quad (0 < s < 1), \tag{2.11}$$

where

$$\left.\begin{array}{l} \phi_r^k(h, s) = \left(\dfrac{ih}{2\pi}\right)^{r+1} A_{rk}(0, s) \quad (k = 0, n), \\[2ex] R = \sigma_n(s) + \rho_n(s) - [\sigma_0(s) + \rho_0(s)] + J(s). \end{array}\right\} \qquad (2.12)$$

and

From (2.5–8) it may be noticed that the correction terms in (2.11) are intended to approximate the two integrals $I_0(s)$ and $I_n(s)$ in the right side of the error lemma (2.1).

If h is suitably small, both $J(s)$ and $\sigma_k(s)$ are of order $O(e^{-2\pi b/h})$ (see (2.3 and 10)) and $\rho_k(s)$ is $O(h^{m_k+2})$ (see (2.5 and 7)), so that the truncation error R in (2.11) may be made as small as we please, provided h and m_k are suitably chosen.

Similar statements are also true of the following *trapezoidal formula with prescribed weighting functions* which we may derive by applying the foregoing analysis to the corollary (1.7), in the form

$$h\left\{\tfrac{1}{2}F(x_0) + \sum_{r=1}^{n-1} F(x_r) + \tfrac{1}{2}F(x_n)\right\} - \int_{x_0}^{x_n} F(x)\,\mathrm{d}x$$

$$= \sum_{r=0}^{m_n} \frac{\tau_r^n(h)}{r!} f_n^{(r)}(x_n) - \sum_{r=0}^{m_0} \frac{\tau_r^0(h)}{r!} f_0^{(r)}(x_0) + R, \qquad (2.13)$$

where

$$\left.\begin{array}{l} \tau_r^k(h) = \left(\dfrac{ih}{2\pi}\right)^{r+1} T_{rk}(0) \quad (k = 0, n), \\[2ex] T_{rk}(\xi) = \displaystyle\int_\xi^\infty \frac{(-)^r w_k(x_k - iht/2\pi) - w_k(x_k + iht/2\pi)}{e^t - 1}\, t^r\,\mathrm{d}t, \end{array}\right\} \qquad (2.14)$$

where

and

$$R = \sigma_n(0) + \rho_n(0) - \{\sigma_0(0) + \rho_0(0)\} + J(0), \qquad (2.15)$$

where $J(0)$, $\rho_k(0)$ and $\sigma_k(0)$ are obtained by writing $s = 0$ in (2.3), (2.7) and (2.10) respectively.

Thus we have not only established that, for suitably chosen weighting functions, formulae of the forms (2.11 and 13) exist, but at the same time we have derived expressions for the associated remainder terms and, more importantly, we have obtained the formulae (2.9 and 14) which express the relevant correction coefficients explicitly in terms of a class of definite integrals and provide us with a direct method for evaluating these coefficients.

In the remaining sections of this chapter we proceed to use the formulae of this section to obtain the respective special formulae for several important weighting functions. The remainder, in each case, is denoted by R, and its explicit form may be obtained from (2.12) with the help of the information available in the relevant section.

In these following sections the priority is always given to obtaining the general rectangular formula (2.11) from which it is usually possible to derive the trapezoidal formula (2.13) as a special case.

In this connection it may be noted that if, after evaluating the integrals $A_{rk}(0, s)$ of the rectangular formula (2.11) explicitly in terms of the variable s, where $0 < s < 1$, it is found that

$$\lim_{s \to 0} A_{rk}(0, s) = A_{rk}(0, 0) \quad \text{and} \quad \lim_{s \to 1} A_{rk}(0, s) = A_{rk}(0, 1)$$

exist, then letting s tend to 0 and 1 in (2.11) and comparing the two resultant formulae with (2.13) we may, since $f_k(x)$ is arbitrary, deduce the interesting continuity relations

$$\left. \begin{aligned} T_{0k}(0) &= \tfrac{1}{2}\{A_{0k}(0, 0) + A_{0k}(0, 1)\}, \\ A_{0k}(0, 1) - A_{0k}(0, 0) &= -2\pi i w_k(x_k) \\ T_{rk}(0) &= A_{rk}(0, 0) = A_{rk}(0, 1) \ (r > 0), \end{aligned} \right\} \tag{2.16}$$

and

valid for $k = 0$ or n.

$$3. \quad \int_{x_0}^{x_n} e^{px} f(x) \, dx$$

Here, corresponding to (2.4), we have

$$f_k(z) = f(z) \quad \text{and} \quad w_k(z) = e^{pz} \quad (k = 0, n), \tag{3.1}$$

so that, from (2.9),

$$A_{rk}(\xi, s) = e^{px_k} \int_{\xi}^{\infty} \left\{ \frac{(-)^r e^{-iqt/2\pi}}{e^{-2\pi is} e^t - 1} - \frac{e^{iqt/2\pi}}{e^{2\pi is} e^t - 1} \right\} t^r \, dt,$$

$$= e^{px_k} I_r(s, q, \xi), \tag{3.2}$$

say, where

$$q = ph. \tag{3.3}$$

Then, by differentiating $I_r(s, q, \xi)$ u times partially with respect to q, we obtain from (3.2)

$$(-2\pi i)^u \frac{\partial^u}{\partial q^u} I_r(s, q, \xi) = I_{r+u}(s, q, \xi) \tag{3.4}$$

or, in particular, for $r = 0$ and writing r for u,

$$I_r(s, q, \xi) = (-2\pi i)^r \frac{\partial^r}{\partial q^r} I_0(s, q, \xi). \tag{3.5}$$

2

Hence, using (3.2 and 5) in (2.11) and writing

$$I_0(s, q, 0) = -2\pi i \phi(s, q), \tag{3.6}$$

we may express the *general rectangular formula with the exponential weighting function* in the form

$$h \sum_{r=0}^{n-1} e^{p(x_r + sh)} f(x_r + sh) - \int_{x_0}^{x_n} e^{px} f(x) \, dx$$

$$= \sum_{r=0}^{m_n} h^{r+1} \frac{L_r(s, q)}{r!} e^{px_n} f^{(r)}(x_n)$$

$$- \sum_{r=0}^{m_0} h^{r+1} \frac{L_r(s, q)}{r!} e^{px_0} f^{(r)}(x_0) + R(s, q), \tag{3.7}$$

where

$$L_r(s, q) = \partial^r / \partial q^r \, \phi(s, q). \tag{3.8}$$

In order to obtain $\phi(s, q)$ we first observe that, for $m_0 = m_n = m$, the remainder $R(s, q)$ in (3.7) is zero whenever $f(x) = P_m(x)$, a polynomial of degree m in x.

Thus, for $f(x) = P_0(x) = 1$ with $x_0 = 0$ and $x_n = h$, we obtain from (3.7)

$$h e^{phs} - \frac{e^{ph} - 1}{p} = h\phi(s, q) (e^{ph} - 1),$$

which yields, on writing $p = q/h$,

$$\phi(s, q) = \frac{e^{sq}}{e^q - 1} - \frac{1}{q}. \tag{3.9}$$

Hence also

$$L_r(s, q) = e^{sq} D_r(s, q), \tag{3.10}$$

where $D_r(s, q)$ has the generating function

$$\frac{e^{st}}{e^q e^t - 1} - \frac{e^{-sq}}{q + t} = \sum_{r=0}^{\infty} \frac{t^r}{r!} D_r(s, q). \tag{3.11}$$

We should note here that the functions $D_r(s, q)$ defined above possess certain important finite difference and other properties which, together with the similar properties of the allied polynomials $C_r(s, q)$, are developed later in Chapter II, where, based on these properties, alternative methods for deriving the general formula (3.7) are also given, including a method of integration by parts which gives rise to a different and, perhaps, more convenient expression for the remainder $R(s, q)$ when $m_n = m_0 = m$ in the formula. On account of the relationship (3.10), the formulae for the calculation of $D_r(s, q)$ and $C_r(s, q)$ given in Chapter II may also be used for the computation of the coefficients $L_r(s, q)$ as well as the coefficients $\lambda_r(q)$

of (3.14). It is also established in Chapter II that these coefficients exist for all q and s, except for $q = 2m\pi i$, $m = \pm 1, \pm 2, \pm 3, \ldots$.

We should also note that the expression (3.10) for $L_r(s, q)$, which was obtained in the present section by simple analytical techniques, may also be derived by the direct evaluation of the integral representation (3.2) with the help of the properties of the higher transcendental Φ-function. This method is described in Appendix A.

Yet another approach is to notice that the form (3.7), without the relationship (3.8), is a direct consequence of (3.2).

If we then take $m_n = m_0 = m$ and $f(x) = e^x$, (3.7) reduces to

$$e^{sq}\left\{\frac{e^{sh}}{e^q\,e^h - 1} - \frac{e^{-sq}}{q+h}\right\} = \sum_{r=0}^{m} h^r \frac{L_r(s, q)}{r!} + R(s, q),$$

which establishes (3.10).

In numerical computations the formulae more commonly in use are those which correspond either to the mid-ordinate sum or to the trapezoidal sum. In these cases the coefficients are also easier to compute (cf. Chapter II, Section 4).

The formula corresponding to the *mid-ordinate sum* is easily obtained by writing $s = \frac{1}{2}$ in the general formula (3.7):

$$h\sum_{r=0}^{n-1} e^{p(x_r+h/2)} f\left(x_r + \frac{h}{2}\right) - \int_{x_0}^{x_n} e^{px} f(x)\,dx$$

$$= \sum_{r=0}^{m_n} h^{r+1} \frac{L_r(\frac{1}{2}, q)}{r!}\, e^{px_n} f^{(r)}(x_n)$$

$$- \sum_{r=0}^{m_0} h^{r+1} \frac{L_r(\frac{1}{2}, q)}{r!}\, e^{px_0} f^{(r)}(x_0) + R(\tfrac{1}{2}, q). \qquad (3.12)$$

On the other hand, the *trapezoidal sum formula* corresponds to writing $s = 0$ (or $s = 1$) in (3.7), when, using the notation

$$\lambda_r(q) \quad \begin{cases} = L_0(0, q) + \frac{1}{2} & (r = 0), \\[2mm] = L_r(0, q) & (r \geqslant 1), \end{cases} \qquad (3.13)$$

we obtain

$$h\left\{\tfrac{1}{2} e^{px_0} f(x_0) + \sum_{r=1}^{n-1} e^{px_r} f(x_r) + \tfrac{1}{2} e^{px_n} f(x_n)\right\} - \int_{x_0}^{x_n} e^{px} f(x)\,dx$$

$$= \sum_{r=0}^{m_n} h^{r+1} \frac{\lambda_r(q)}{r!}\, e^{px_n} f^{(r)}(x_n)$$

$$- \sum_{r=0}^{m_0} h^{r+1} \frac{\lambda_r(q)}{r!}\, e^{px_0} f^{(r)}(x_0) + R(0, q). \qquad (3.14)$$

In Appendix C we give interpolatable tables of $\lambda_r(\pm q)$ $(0 \leqslant r \leqslant 11)$ for $q = u$ and $q = iv$ and for $u, v = 0 \ (0.02) \ 1$.

4. *The weighting functions* $\cosh(\alpha x)$, $\sinh(\alpha x)$, $\cos(\beta x)$ *and* $\sin(\beta x)$

A weighting function $w(x)$ which is a polynomial in $\cosh(a_j x)$, $\sinh(b_j x)$, $\cos(\alpha_j x)$, $\sin(\beta_j x)$ and $e^{c_j x}$, where a_j, b_j, α_j, β_j and c_j $(j = 1, 2, 3, ...)$ are constants, may be expressed as a linear combination of weighting functions of the form $w_j(x) = e^{p_j x}$ $(j = 1, 2, 3, ...)$. The rectangular formula for any such weighting function may therefore be derived by forming exactly the same linear combination of the corresponding set of formulae obtained by writing p_j $(j = 1, 2, 3, ...)$ for p in (3.7). The mid-ordinate and the trapezoidal formulae may similarly be derived from (3.12) and (3.14) respectively.

Thus, for example, if we take half the sum and half the difference of the two formulae which correspond to putting $p = \alpha$ and $p = -\alpha$ in (3.7), we obtain the rectangular formulae corresponding to the weighting functions $\cosh(\alpha x)$ and $\sinh(\alpha x)$ respectively, while the same procedure with $i\beta$ in place of α results in the pair of rectangular formulae for the weighting functions $\cos(\beta x)$ and $\sin(\beta x)$.

Adopting the notation

$$\alpha h = u, \quad \beta h = v, \tag{4.1}$$

these *rectangular formulae* may be expressed in the form

$$h \sum_{r=0}^{n-1} w(x_r + sh) f(x_r + sh) - \int_{x_0}^{x_n} w(x) f(x) \, dx$$

$$= \sum_{r=0}^{m_n} h^{r+1} \frac{G_r(x_n)}{r!} f^{(r)}(x_n) - \sum_{r=0}^{m_0} h^{r+1} \frac{G_r(x_0)}{r!} f^{(r)}(x_0) + R, \tag{4.2}$$

where the coefficient $G_r(x_k)$ $(k = 0, n)$ and the remainder R, corresponding to each weighting function, are given by the following array:

$w(x)$	$G_r(x_k)$	R
$\cosh(\alpha x)$	$\frac{1}{2}\{L_r(s, u)\,e^{\alpha x_k} + L_r(s, -u)\,e^{-\alpha x_k}\}$	$\frac{1}{2}\{R(s, u) + R(s, -u)\}$
$\sinh(\alpha x)$	$\frac{1}{2}\{L_r(s, u)\,e^{\alpha x_k} - L_r(s, -u)\,e^{-\alpha x_k}\}$	$\frac{1}{2}\{R(s, u) - R(s, -u)\}$
$\cos(\beta x)$	$\frac{1}{2}\{L_r(s, iv)\,e^{i\beta x_k} + L_r(s, -iv)\,e^{-i\beta x_k}\}$	$\frac{1}{2}\{R(s, iv) + R(s, -iv)\}$
$\sin(\beta x)$	$\frac{1}{2i}\{L_r(s, iv)\,e^{i\beta x_k} - L_r(s, -iv)\,e^{-i\beta x_k}\}$	$\frac{1}{2i}\{R(s, iv) - R(s, -iv)\}$

$$\tag{4.3}$$

It is useful to notice that, by virtue of (3.8–9) and (3.11), both the coefficients $L_r(\tfrac{1}{2}, q)$ and $\lambda_r(q)$, corresponding, respectively, to the mid-ordinate and the trapezoidal sums, are odd functions of q when r is even, and even functions of q when r is odd; so that in these cases the coefficients in (4.2) are considerably simplified.

Thus the formulae corresponding to the *mid-ordinate sum* may be written in the form

$$h\sum_{r=0}^{n-1} w\left(x_r+\frac{h}{2}\right) f\left(x_r+\frac{h}{2}\right) - \int_{x_0}^{x_n} w(x) f(x)\,\mathrm{d}x$$

$$= \sum_{r=0}^{m_n} h^{r+1} \frac{G_r(x_n)}{r!} f^{(r)}(x_n) - \sum_{r=0}^{m_0} h^{r+1} \frac{G_r(x_0)}{r!} f^{(r)}(x_0) + R, \qquad (4.4)$$

where the coefficient $G_r(x_k)$ ($k = 0, n$) and the remainder R, corresponding to each weighting function, are given by the array

$$
\begin{array}{lll}
w(x) & G_r(x_k) & R \\[4pt]
\cosh(\alpha x) & L_r(\tfrac{1}{2}, u)\, H_r(\alpha x_k) & \tfrac{1}{2}\{R(\tfrac{1}{2}, u) + R(\tfrac{1}{2}, -u)\} \\[4pt]
\sinh(\alpha x) & L_r(\tfrac{1}{2}, u)\, H_{r+1}(\alpha x_k) & \tfrac{1}{2}\{R(\tfrac{1}{2}, u) - R(\tfrac{1}{2}, -u)\} \\[4pt]
\cos(\beta x) & L_r(\tfrac{1}{2}, iv)\, H_r(i\beta x_k) & \tfrac{1}{2}\{R(\tfrac{1}{2}, iv) + R(\tfrac{1}{2}, -iv)\} \\[4pt]
\sin(\beta x) & -iL_r(\tfrac{1}{2}, iv)\, H_{r+1}(i\beta x_k) & \dfrac{1}{2i}\{R(\tfrac{1}{2}, iv) - R(\tfrac{1}{2}, -iv)\}
\end{array}
\right\} \quad (4.5)
$$

where

$$
H_r(\alpha x_k) = \begin{cases} \sinh(\alpha x_k) & (r\ \text{even}) \\ \cosh(\alpha x_k) & (r\ \text{odd}), \end{cases} \quad
H_r(i\beta x_k) = \begin{cases} i\sin(\beta x_k) & (r\ \text{even}) \\ \cos(\beta x_k) & (r\ \text{odd}). \end{cases} \quad (4.6)
$$

Similarly, the formulae corresponding to the *trapezoidal sum* assume the form

$$h\left\{\tfrac{1}{2}w(x_0) f(x_0) + \sum_{r=1}^{n-1} w(x_r) f(x_r) + \tfrac{1}{2} w(x_n) f(x_n)\right\} - \int_{x_0}^{x_n} w(x) f(x)\,\mathrm{d}x$$

$$= \sum_{r=0}^{m_n} h^{r+1} \frac{G_r(x_n)}{r!} f^{(r)}(x_n) - \sum_{r=0}^{m_0} h^{r+1} \frac{G_r(x_0)}{r!} f^{(r)}(x_0) + R, \qquad (4.7)$$

where now the appropriate array for the coefficient $G_r(x_k)$ and the remainder R is as follows:

$$
\begin{array}{lll}
w(x) & G_r(x_k) & R \\[4pt]
\cosh(\alpha x) & \lambda_r(u)\, H_r(\alpha x_k) & \tfrac{1}{2}\{R(0, u) + R(0, -u)\} \\[4pt]
\sinh(\alpha x) & \lambda_r(u)\, H_{r+1}(\alpha x_k) & \tfrac{1}{2}\{R(0, u) - R(0, -u)\} \\[4pt]
\cos(\beta x) & \lambda_r(iv)\, H_r(i\beta x_k) & \tfrac{1}{2}\{R(0, iv) + R(0, -iv)\} \\[4pt]
\sin(\beta x) & -i\lambda_r(iv)\, H_{r+1}(i\beta x_k) & \dfrac{1}{2i}\{R(0, iv) - R(0, -iv)\}
\end{array}
\right\} \quad (4.8)
$$

where, as before, $H_r(\alpha x_k)$ and $H_r(i\beta x_k)$ are given by (4.6).

It may be noticed that in each of the formulae of the present section the remainder term R has been expressed in terms of the remainder $R(s,q)$ ($s = s$, $\tfrac{1}{2}$ or 0 and $q = \pm \alpha$ or $\pm i\beta$) of (3.7), the complex variable expression for which may be obtained from (2.12). For $m_n = m_0 = m$, a different expression for $R(s,q)$ is obtained in Chapter II where, based on this expression, some upper bounds for the remainder terms in the formulae of Sections 3 and 4 are also obtained, followed by discussions regarding the use of these formulae and some illustrative numerical examples.

5. *The Fourier coefficients*

We should note here that if the above formulae (4.4) and (4.7) are used for the evaluation of the *Fourier coefficients*

$$a_j = \int_a^b \sin\left\{\frac{2\pi j}{b-a}\left(x - \frac{b+a}{2}\right)\right\} f(x)\, \mathrm{d}x = \int_{-(b-a)/2}^{+(b-a)/2} \sin\left(\frac{2\pi j x}{b-a}\right) f\left(x + \frac{b+a}{2}\right) \mathrm{d}x$$

and

$$b_j = \int_a^b \cos\left\{\frac{2\pi j}{b-a}\left(x - \frac{b+a}{2}\right)\right\} f(x)\, \mathrm{d}x = \int_{-(b-a)/2}^{+(b-a)/2} \cos\left(\frac{2\pi j x}{b-a}\right) f\left(x + \frac{b+a}{2}\right) \mathrm{d}x,$$

where j is an integer, then, in each case, we have $\beta = 2\pi j/(b-a)$, $x_0 = -(b-a)/2$ and $x_n = (b-a)/2$; so that $\sin(\beta x_k) = 0$ for both $k = 0$ and $k = n$. Consequently every alternate $H_r(i\beta x_k)$ will vanish leading to some useful saving in computation.

We should also note that it is possible to obtain special formulae for the calculation of the correction coefficients for the integration formulae in these cases (cf. Chapter II, Section 4).

$$6. \quad \int_0^{x_n} \frac{\sin(\beta x)}{x} f(x)\, \mathrm{d}x$$

Here

$$w_0(z) = \frac{\sin(\beta z)}{z} \quad \text{and} \quad w_n(z) = \sin(\beta z),$$

so that

$$f_0(z) = f(z) \quad \text{and} \quad f_n(z) = \frac{f(z)}{z}.$$

$$(6.1)$$

Hence, from (2.9),

$$A_{r0}(\xi, s) = \frac{2\pi}{h} \int_\xi^\infty \left\{\frac{(-1)^r}{e^{-2\pi i s}\, e^t - 1} - \frac{1}{e^{2\pi i s}\, e^t - 1}\right\} \sinh\left(\frac{vt}{2\pi}\right) t^{r-1}\, \mathrm{d}t, \qquad (6.2)$$

where, as before, $v = \beta h$.

Let us consider the two cases $s = \frac{1}{2}$ and $s = 0$ separately.

When $s = \frac{1}{2}$, (6.2) reduces to

$$A_{r0}(\xi, \tfrac{1}{2}) = \{(-1)^r - 1\}\frac{2\pi}{h}\int_\xi^\infty \frac{\sinh(vt/2\pi)}{-(e^t + 1)} t^{r-1}\, \mathrm{d}t, \tag{6.3}$$

so that

$$A_{r0}(\xi, \tfrac{1}{2}) = \begin{cases} 0 & (r\ \text{even}) \\[2mm] \dfrac{2\pi}{h}\displaystyle\int_\xi^\infty \dfrac{e^{vt/2\pi} - e^{-vt/2\pi}}{e^t + 1} t^{r-1}\,\mathrm{d}t & (r\ \text{odd}). \end{cases} \tag{6.4}$$

It may also be shown that (cf. (3.2) with $s = \frac{1}{2}$)

$$\int_0^\infty \frac{e^{vt/2\pi} - (-)^r e^{-vt/2\pi}}{e^t + 1} t^r\, \mathrm{d}t = (2\pi i)^{r+1} L_r(\tfrac{1}{2}, iv), \tag{6.5}$$

where $L_r(\tfrac{1}{2}, iv)$ is given by (3.8–9) or (3.10).

Therefore

$$A_{r0}(0, \tfrac{1}{2}) = \begin{cases} 0 & (r\ \text{even}), \\[2mm] \dfrac{(i)^r (2\pi)^{r+1}}{h} L_{r-1}(\tfrac{1}{2}, iv) & (r\ \text{odd}). \end{cases} \tag{6.6}$$

Also the correction terms for the upper limit x_n in this case will be identical to the correction terms for the same upper limit of (4.4) with $w(x) = \sin(\beta x)$.

Hence we obtain, using (6.3) in (2.11) for $s = \frac{1}{2}$, the *mid-ordinate formula*

$$h\sum_{r=0}^{n-1} \frac{\sin\{\beta(x_r + h/2)\}}{x_r + h/2} f\left(x_r + \frac{h}{2}\right) - \int_0^{x_n} \frac{\sin(\beta x)}{x} f(x)\, \mathrm{d}x$$

$$= \sum_{r=0}^{m_n} h^{r+1} \frac{-iL_r(\tfrac{1}{2}, iv)\, H_{r+1}(i\beta x_n)}{r!} f^{(r)}(x_n)$$

$$- \sum_{r=0}^{m_0} h^{2r+1} \frac{-iL_{2r}(\tfrac{1}{2}, iv)}{(2r+1)!} f^{(2r+1)}(0) + R, \tag{6.7}$$

where $H_r(i\beta x_n)$ is given by (4.6).

When $s = 0$, (6.2) reduces to (see (2.14))

$$T_{r0}(\xi) = \{(-1)^r - 1\}\frac{2\pi}{h}\int^\infty \frac{\sinh(vt/2\pi)}{e^t - 1} t^{r-1}\, \mathrm{d}t, \tag{6.8}$$

so that

$$T_{r0}(\xi) = \begin{cases} 0 & (r\ \text{even}) \\[2mm] -\dfrac{2\pi}{h}\displaystyle\int_\xi^\infty \dfrac{e^{vt/2\pi} - e^{-vt/2\pi}}{e^t - 1} t^{r-1}\,\mathrm{d}t & (r\ \text{odd}). \end{cases} \tag{6.9}$$

It may also be shown that (cf. (3.2) with $s = 0$)

$$\int_0^\infty \frac{e^{vt/2\pi} - (-)^r e^{-vt/2\pi}}{e^t - 1} \, t^r \, dt = -(2\pi i)^{r+1} \lambda_r(iv), \tag{6.10}$$

where $\lambda_r(iv)$ is given by (3.13).

Therefore

$$T_{r0}(0) = \begin{cases} 0 & (r \text{ even}), \\[2ex] \dfrac{(i)^r (2\pi)^{r+1}}{h} \lambda_{r-1}(iv) & (r \text{ odd}). \end{cases} \tag{6.11}$$

We also have that the correction terms for the upper limit x_n in this case will be identical to the correction terms for the same upper limit of (4.7) with $w(x) = \sin(\beta x)$.

Hence we obtain, using (6.11) in (2.13), the *trapezoidal formula*

$$h\left\{ \frac{\beta}{2} f(0) + \sum_{r=1}^{n-1} \frac{\sin(\beta rh)}{rh} f(rh) + \frac{1}{2} \frac{\sin(\beta nh)}{nh} f(nh) \right\} - \int_{x_0}^{x_n} \frac{\sin(\beta x)}{x} f(x) \, dx$$

$$= \sum_{r=0}^{m_n} h^{r+1} \frac{-i\lambda_r(iv) \, H_{r+1}(i\beta x_n)}{r!} \frac{d^r}{dx^r} \left\{ \frac{f(x)}{x} \right\} \Bigg|_{x=x_n}$$

$$- \sum_{r=0}^{m_0} h^{2r+1} \frac{-i\lambda_{2r}(iv)}{(2r+1)!} f^{(2r+1)}(0) + R, \tag{6.12}$$

where $H_r(i\beta x_n)$ is given by (4.6).

$$7. \quad \int_{x_0}^{x_n} (x - x_0)^{\alpha_0} f(x) \, dx$$

Here

$$\left. \begin{aligned} w_0(z) &= (z - x_0)^{\alpha_0}, & f_0(z) &= f(z), \\ w_n(z) &= 1, & f_n(z) &= (z - x_0)^{\alpha_0} f(z), \end{aligned} \right\} \tag{7.1}$$

so that

$$w_0\left(x_0 + \frac{iht}{2\pi} \right) = \left(\frac{ht}{2\pi} \right)^{\alpha_0} e^{\alpha_0 \pi i/2} \tag{7.2}$$

and

$$A_{r0}(\xi, s) = \int_\xi^\infty \left(\frac{h}{2\pi} \right)^{\alpha_0} \left\{ \frac{(-)^r e^{-\alpha_0 \pi i/2}}{e^{-2\pi is} e^t - 1} - \frac{e^{\alpha_0 \pi i/2}}{e^{2\pi is} e^t - 1} \right\} t^{r+\alpha_0} \, dt. \tag{7.3}$$

We also know that (*Higher transcendental functions*, vol. I), for $\operatorname{Re}\{z\} > 1$ and $0 < s \leqslant 1$ (or $0 < s < 1$, $\operatorname{Re}\{z\} > 0$),

$$e^{z\pi i/2} \zeta_s(z) + e^{z\pi i/2} \zeta_{-s}(z) = \frac{(2\pi)^z}{\Gamma(z)} \zeta(1-z, s), \tag{7.4}$$

where

$$\zeta_s(z) = \sum_{n=1}^{\infty} \frac{e^{2n\pi i s}}{n^z} = \frac{1}{\Gamma(z)} \int_0^{\infty} \frac{t^{z-1}}{e^{-2\pi i s}\, e^t - 1}\, dt, \tag{7.5}$$

and where $\zeta(z, s)$ is the well-known generalized Riemann zeta function.

Therefore, using (7.4) in (7.3), we obtain

$$A_{r0}(0, s) = (i)^{1-r} h^{\alpha_0} (2\pi)^{r+1} \zeta\{-(\alpha_0 + r), s\}. \tag{7.6}$$

Taking $\alpha_0 = 0$ in (7.6) we also obtain

$$A_{rn}(0, s) = (i)^{1-r}(2\pi)^{r+1}\, \zeta(-r, s) = -\frac{(i)^{1-r}(2\pi)^{r+1}}{(r+1)} B_{r+1}(s), \tag{7.7}$$

where $B_{r+1}(s)$ is the Bernoulli polynomial of order $(r+1)$.

Hence using (7.6) and (7.7) in (2.11) we obtain, finally, the *rectangular formula*

$$h \sum_{r=0}^{n-1} (rh + sh)^{\alpha_0} f(x_r + sh) - \int_{x_0}^{x_n} (x - x_0)^{\alpha_0} f(x)\, dx$$

$$= \sum_{r=0}^{m_n} h^{r+1} \frac{B_{r+1}(s)}{(r+1)!} f_n^{(r)}(x_n) + \sum_{r=0}^{m_0} h^{r+1+\alpha_0} \frac{\zeta\{-(\alpha_0 + r), s\}}{r!} f_0^{(r)}(x_0) + R. \tag{7.8}$$

The respective formulae for the *mid-ordinate sum* and the *trapezoidal sum* (the latter not in its standard form) are easily obtained by taking, in turn, $s = \frac{1}{2}$ and $s = 1$ in equation (7.8), which is seen to hold for

$$0 < s \leqslant 1, \quad \mathrm{Re}\{\alpha_0\} > -1. \tag{7.9}$$

In these cases we also have the useful relations

$$\zeta(z, 1) = \zeta(z) \quad \text{and} \quad \zeta(z, \tfrac{1}{2}) = (2^z - 1)\, \zeta(z), \tag{7.10}$$

which facilitate the computation of the coefficients, since $\zeta(z)$, the Riemann zeta function, is well tabulated (see, e.g., Dwight, 1958).

$$8. \quad \int_{x_0}^{x_n} (x - x_0)^{\alpha_0}(x_n - x)^{\alpha_n} f(x)\, dx$$

We should note here that formula (7.8) is adequate for the evaluation of an integral of the more general form given above, since this integral is easily expressed as the sum of two integrals, each of which is of the form treated by this formula.

Thus, or otherwise by a direct application of (2.11), we obtain, writing

$$\left. \begin{array}{l} F(z) = (z - x_0)^{\alpha_0}(x_n - z)^{\alpha_n} f(z), \quad f_0(z) = (x_n - z)^{\alpha_n} f(z) \\[2mm] f_n(z) = (z - x_0)^{\alpha_0} f(z), \end{array} \right\} \tag{8.1}$$

and

the formulae

$$h \sum_{r=0}^{n-1} F\left(x_r + \frac{h}{2}\right) - \int_{x_0}^{x_n} F(x)\,\mathrm{d}x$$

$$= \sum_{r=0}^{m_n} (-)^r\, h^{r+1+\alpha_n} \frac{\{2^{-(r+\alpha_n)} - 1\}\,\zeta\{-(r+\alpha_n)\}}{r!} f_n^{(r)}(x_n)$$

$$+ \sum_{r=0}^{m_0} h^{r+1+\alpha_0} \frac{\{2^{-(r+\alpha_0)} - 1\}\,\zeta\{-(r+\alpha_0)\}}{r!} f_0^{(r)}(x_0) + R \quad (8.2)$$

for the *mid-ordinate sum*, and

$$h \sum_{r=1}^{n-1} F(x_r) - \int_{x_0}^{x_n} F(x)\,\mathrm{d}x = \sum_{r=0}^{m_n} (-)^r\, h^{r+1+\alpha_n} \frac{\zeta\{-(r+\alpha_n)\}}{r!} f_n^{(r)}(x_n)$$

$$+ \sum_{r=0}^{m_0} h^{r+1+\alpha_0} \frac{\zeta\{-(r+\alpha_0)\}}{r!} f_0^{(r)}(x_0) + R \quad (8.3)$$

for the *trapezoidal sum*.

In Appendix C we give values of $\zeta\{-(r+\alpha)\}$ for $r = 0\,(1)\,8$ and for $\alpha = \pm\frac{1}{4},\ \pm\frac{1}{3},\ \pm\frac{1}{2},\ \pm\frac{2}{3}$ and $\pm\frac{3}{4}$ to eight figures.

CHAPTER II

THE C-POLYNOMIALS AND THE D-FUNCTIONS

Introduction

In Chapter I we used complex variable methods to derive general rectangular formulae with prescribed weighting functions. In the present chapter we show that the class of polynomials $C_r(z, q)$ (defined by (1.1)) and the class of functions $D_r(z, q)$ (defined by (3.1)), which are closely related to the correction coefficients for the exponential weighting function, possess several important finite difference and other properties with applications to summation of series problems, finite difference equations, and numerical integration involving the exponential and allied weighting functions.

In Chapter I the general rectangular formulae for the exponential and allied weighting functions were obtained by complex variable methods. In Section 7 of this chapter we give a different treatment based on the properties of the D-functions. The derivation yields a different and, perhaps, more tractable form for the remainder term which is then analysed to obtain useful estimates and bounds for the truncation error. This is followed by discussions regarding the use of these formulae, comparison with other quadrature formulae and some illustrative numerical examples.

In Section 10 the properties of the C-polynomials are used to obtain the summation formula (10.4) for sums of the form $\sum e^{px_r} f(x_r)$, the operational development of which for the special case $s = 0$, but without the form for the remainder term, was given by Boole in 1860. The special case of an alternating series to which (10.4) reduces for $q = ph = i\pi$ has been treated in Nörlund (1924) with the help of the properties of the Euler polynomials to which the C-polynomials reduce for $q = i\pi$.

In Sections 4 and 6 we derive several formulae for the computation of $C_r(s, q)$ and $D_r(s, q)$. On account of the relationship (3.10) of Chapter I, these formulae are also useful for the computation of the coefficients $L_r(s, q)$ and $\lambda_r(q)$ which appear in the formulae for the exponential and allied weighting functions (see Sections 3–6 of Chapter I).

1. *The C-polynomials*

The C-polynomials and the D-functions may be defined in various ways—some of their distinctive properties may as well be used as definitions—the generating function definitions adopted here arise naturally from the relationship (3.10) of Chapter I. In this connection we may notice that, although the C-polynomials are special cases of the Φ-function (a proof of this is given in Appendix A), a simple treatment based on the generating function definition (1.1) yields the necessary properties without recourse to the analytical properties of the Φ-function.

Let the relation

$$\frac{e^{zt}}{e^{q}e^{t}-1} = \sum_{r=0}^{\infty}\frac{t^{r}}{r!}C_{r}(z,q) \quad (q\neq 0, \pm 2m\pi i) \tag{1.1}$$

define the polynomial $C_r(z,q)$ of degree r.

For brevity, we shall also use the notation

$$C_r(z,q) \equiv C_r(z)$$

when the common value of q will be clear from the context.

When $z = 0$, (1.1) reduces to

$$\frac{1}{e^{q}e^{t}-1} = \sum_{r=0}^{\infty}\frac{t^{r}}{r!}C_{r}(0,q), \tag{1.2}$$

which defines, for a given value of the parameter q, an associated set of numbers $C_r(0,q) \equiv C_r(0)$.

From (1.1) it follows that†

$$\left. \begin{aligned} C_r(y+z) &= \sum_{j=0}^{r}\binom{r}{j}C_j(y)\,z^{r-j} \\ &= \{C(y)+z\}^{r}, \end{aligned} \right\} \tag{1.3}$$

symbolically, where, after expansion, each $\{C(y)\}^j$ $(j = 0, 1, \dots, r)$ is to be replaced by $C_j(y)$.

On writing $y = 0$, (1.3) yields the important formula

$$C_r(z) = \{C(0)+z\}^{r}. \tag{1.4}$$

The relationship (1.4) shows that $C_r(z)$ is actually a polynomial of degree r in z, and that these polynomials $C_r(z)$ are completely determined by (1.4) together with the numbers $C_r(0)$ defined by (1.2).

---

† The following relationship (1.3) as well as the relations (1.4–6), which follow from (1.3), are in fact satisfied by a more general class of functions, including the Bernoulli polynomials (see, e.g., Milne-Thomson, 1951). The results (1.8–17) obtained later, however, are characteristic of the C-polynomials only.

It also follows from (1.4) that

$$\frac{\mathrm{d}}{\mathrm{d}z} C_r(z) = r C_{r-1}(z) \tag{1.5}$$

and that

$$\int_a^z C_r(t)\,\mathrm{d}t = \frac{C_{r+1}(z) - C_{r+1}(a)}{r+1}. \tag{1.6}$$

Furthermore, if Δ_q be the weighted difference operator

$$\Delta_q = \mathrm{e}^q E - 1, \tag{1.7}$$

then the operation of Δ_q on both sides of (1.1) yields

$$\sum_{r=0}^{\infty} \frac{t^r}{r!} \Delta_q C_r(z) = (\mathrm{e}^q \mathrm{e}^t - 1) \frac{\mathrm{e}^{zt}}{\mathrm{e}^q \mathrm{e}^t - 1}$$

$$= \mathrm{e}^{zt}$$

$$= \sum_{r=0}^{\infty} \frac{t^r z^r}{r!},$$

whence, equating the coefficients of t^r, we obtain the weighted difference equation

$$\Delta_q C_r(z) = z^r, \tag{1.8}$$

which, for $z = 0$, reduces to the relationship

$$\left.\begin{array}{l} C_r(1) = \mathrm{e}^{-q} C_r(0) \quad (r > 0), \\ C_0(1) = C_0(0) = 1/(\mathrm{e}^q - 1). \end{array}\right\} \tag{1.9}$$

It may also be shown that

$$\mathrm{e}^q C_r(1 - z, q) = (-)^{r+1} C_r(z, -q), \tag{1.10}$$

which is the 'complementary argument theorem' for the C-polynomials.

Using (1.8) in (1.10) above, we obtain the relationship

$$C_r(z, -q) = (-)^{r+1} C_r(-z, q) - z^r. \tag{1.11}$$

The relations (1.10) and (1.11) show that both $\mathrm{e}^{q/2} C_r(\tfrac{1}{2}, q)$ and $C_r(0, q)$ are odd functions of q when r is even and even functions of q when r is odd, except for $C_0(0, q)$ in which case $C_0(0, q) + \tfrac{1}{2}$ is odd in q.

It may also be shown that

$$\frac{C_r(mz, q/m)}{m^r} = \sum_{j=0}^{m-1} \mathrm{e}^{jq/m} C_r(z + j/m, q), \tag{1.12}$$

whence, taking $z = 0$ and $m = 2$, we may obtain

$$\frac{C_r(0, q/2)}{2^r} = C_r(0, q) + \mathrm{e}^{q/2} C_r(\tfrac{1}{2}, q), \tag{1.13}$$

which expresses $C_r(\tfrac{1}{2}, q)$ in terms of $C_r(0, q)$ and $C_r(0, q/2)$.

Again, if m and ν are integral, positive or negative, ν/m not an integer, we have, since

$$\sum_{r=0}^{\infty} \frac{t^r}{r!} B_r(s) = \frac{t\,\mathrm{e}^{st}}{\mathrm{e}^t - 1},$$

$$\sum_{r=0}^{\infty} \frac{t^r}{r!} \sum_{j=0}^{m-1} \mathrm{e}^{2\pi i \nu j/m} B_r\left(\frac{z+j}{m}\right) = \sum_{j=0}^{m-1} \frac{t\,\mathrm{e}^{2\pi i \nu j/m}\,\mathrm{e}^{(z+j)t/m}}{\mathrm{e}^t - 1}$$

$$= \frac{t\,\mathrm{e}^{zt/m}}{\mathrm{e}^{2\pi i \nu/m}\,\mathrm{e}^{t/m} - 1}$$

$$= \sum_{r=0}^{\infty} \frac{t^{r+1}}{r!\,m^r} C_r(z, 2\pi i \nu/m),$$

from (1.1).

Hence

$$C_r(z, 2\pi i \nu/m) = \frac{m^r}{r+1} \sum_{j=0}^{m-1} \mathrm{e}^{2\pi i \nu j/m} B_{r+1}\left(\frac{z+j}{m}\right). \tag{1.14}$$

The following results may also be derived from (1.1):

$$\frac{\mathrm{d}^\nu}{\mathrm{d}q^\nu}\{\mathrm{e}^{zq} C_r(z, q)\} = \mathrm{e}^{zq} C_{r+\nu}(z, q), \tag{1.15}$$

$$C_r(z, q + i\pi) = 2^{r+1} C_r(z/2, 2q) - C_r(z, q) \tag{1.16}$$

and

$$C_r(z, \pi i) = -\tfrac{1}{2} E_r(z), \tag{1.17}$$

where $E_r(z)$ denotes the Euler polynomial.

To prove (1.15) we note that from (1.1)

$$\mathrm{e}^{zq} C_r(z, q) = \frac{\mathrm{d}^r}{\mathrm{d}t^r}\left\{\frac{\mathrm{e}^{z(q+t)}}{\mathrm{e}^{q+t} - 1}\right\}\bigg|_{t=0} = \frac{\mathrm{d}^r}{\mathrm{d}q^r}\left\{\frac{\mathrm{e}^{zq}}{\mathrm{e}^q - 1}\right\}, \tag{1.18}$$

which, differentiated ν times with respect to q, yields (1.15); while (1.16) follows from the expansion of both sides of the equality

$$\frac{\mathrm{e}^{zt}}{\mathrm{e}^{(q+i\pi)}\,\mathrm{e}^t - 1} = \frac{2\,\mathrm{e}^{zt}}{\mathrm{e}^{2q}\,\mathrm{e}^{2t} - 1} - \frac{\mathrm{e}^{zt}}{\mathrm{e}^q\,\mathrm{e}^t - 1},$$

by using (1.1), and subsequently equating the coefficients of like powers of t.

The Euler polynomials are defined as (Nörlund, 1924)

$$\frac{2\,\mathrm{e}^{zt}}{\mathrm{e}^t+1} = \sum_{r=0}^{\infty} \frac{t^r}{r!}\, E_r(z)$$

which, when compared to (1.1) with $q = \pi\mathrm{i}$, yields (1.17).

2. *Some applications*

Application to the solution of difference equations

It is important to notice that since by (1.8)

$$\Delta_q\, C_r(z) = z^r,$$

the linear difference equation

$$\Delta_q\, u(z) = P(z), \tag{2.1}$$

where $P(z)$ is a polynomial, has the symbolical polynomial solution

$$u(z) = P\{C(z)\}, \tag{2.2}$$

where, after expansion, each $\{C(z)\}^r$ $(r = 0, 1, 2, \ldots)$ is to be replaced by $C_r(z)$; or, equivalently, the linear difference equation

$$\Delta u(z) = \mathrm{e}^{qz}\, P(z) \tag{2.3}$$

has the symbolical solution

$$u(z) = \mathrm{e}^{qz}\, P\{C(z)\}. \tag{2.4}$$

A summation formula

If we multiply both sides of (1.8) by e^{sq} and write $x+s$ for z, we obtain, by summing both sides of the ensuing equation from $s = 0$ to $s = n-1$, the summation formula

$$\mathrm{e}^{nq}\, C_r(x+n) - C_r(x) = \sum_{s=0}^{n-1} \mathrm{e}^{sq}(x+s)^r. \tag{2.5}$$

For example, let $\mathrm{e}^q = 2$, $x = 0$ and $r = 1$. Then from (2.5)

$$\sum_{s=0}^{n-1} 2^s\, s = 2^n\, C_1(n) - C_1(0)$$

$$= 2^n\left\{n\left(\frac{2}{2-1}-1\right) - \frac{2}{(2-1)^2}\right\} + \frac{2}{(2-1)^2}$$

(from (4.6) and (1.4), since $\mathrm{e}^q = 2$)

$$= n2^n - 2^{n+1} + 2.$$

An expansion theorem for the C-polynomials

It follows from (2.2) that, $P(z)$ denoting an arbitrary polynomial of degree ν and for $z = x + s$,

$$P(x+s) = e^q P\{C(x+s+1)\} - P\{C(x+s)\}$$

$$= e^q P\{x+1+C(s)\} - P\{x+C(s)\}, \qquad (2.6)$$

by (1.3).

Again, by Taylor's theorem,

$$P\{x+C(s)\} = C_0(s)\,P(x) + C_1(s)\,P'(x)$$

$$+ \frac{1}{2!}\,C_2(s)\,P''(x) + \dots + \frac{1}{\nu!}\,C_\nu(s)P^{(\nu)}(x). \qquad (2.7)$$

Hence, substituting in (2.6), we have the expansion

$$P(x+s) = C_0(s)\,\Delta_q P(x) + C_1(s)\,\Delta_q P'(x)$$

$$+ \frac{1}{2!}\,C_2(s)\,\Delta_q P''(x) + \dots + \frac{1}{\nu!}\,C_\nu(s)\,\Delta_q P^{(\nu)}(x). \qquad (2.8)$$

If we put $x = 0$ in (2.8), we have the expansion of $P(s)$ in terms of the C-polynomials.

The general rectangular formula for a polynomial with the exponential weighting function

If we multiply both sides of (2.8) by $e^{q(s+x)}$ and then add to the right side of the ensuing equation the null sum

$$0 = \int_x^{x+1} e^{qt} P(t)\,dt + \sum_{r=0}^{\nu} \left(\frac{-1}{q}\right)^{r+1} e^{qx}\,\Delta_q P^{(r)}(x), \qquad (2.9)$$

we obtain

$$e^{q(s+x)}\,P(x+s)$$

$$= \int_x^{x+1} e^{qt} P(t)\,dt + \sum_{r=0}^{\nu} \frac{e^{sq}}{r!} \left\{C_r(s) + r!\left(\frac{-1}{q}\right)^{r+1} e^{-sq}\right\} \Delta\{e^{qx} P^{(r)}(x)\} \qquad (2.10)$$

$$= \int_x^{x+1} e^{qt} P(t)\,dt + \sum_{r=0}^{\nu} \frac{e^{sq}\,D_r(s,q)}{r!} \Delta\{e^{qx} P^{(r)}(x)\}, \qquad (2.11)$$

which, it is easy to recognize, is the general rectangular formula for a polynomial with the exponential weighting function.

3. *The functions* $D_r(z, q)$

A close inspection of the definition (1.1) shows that the polynomials $C_r(z, q)$ will not tend to the Bernoulli polynomials in the limit as q is made to tend to zero, although the generating function on the left side of (1.1) will in that case be identical to the generating function for the Bernoulli polynomials barring only a factor t.

A class of functions $D_r(z) \equiv D_r(z, q)$ of z and the parameter q which will tend to the Bernoulli polynomials as a limiting form when q tends to zero† may be defined as

$$D_r(z) \equiv D_r(z, q) = C_r(z, q) + r! \left(\frac{-1}{q} \right)^{r+1} e^{-zq}$$

$$(q \neq \pm 2m\pi i, m = 1, 2, 3, \ldots) \tag{3.1}$$

or, in terms of the generating function,

$$\frac{e^{zt}}{e^q e^t - 1} - \frac{e^{-zq}}{q+t} = \sum_{r=0}^{\infty} \frac{t^r}{r!} D_r(z, q). \tag{3.2}$$

Thus defined, the functions $D_r(z)$ retain the important properties of $C_r(z)$ given by equations (1.5) to (1.13) and (1.15), i.e.

$$\frac{d}{dz} D_r(z) = r D_{r-1}(z), \tag{3.3}$$

$$\int_a^z D_r(t) \, dt = \frac{D_{r+1}(z) - D_{r+1}(a)}{r+1}, \tag{3.4}$$

$$\Delta_q D_r(z) = z^r, \tag{3.5}$$

$$D_r(1) = e^{-q} D_r(0) \quad (r > 0), \tag{3.6}$$

$$e^q D_r(1 - z, q) = (-)^{r+1} D_r(z, -q), \tag{3.7}$$

$$D_r(z, -q) = (-)^{r+1} D_r(-z, q) - z^r, \tag{3.8}$$

$$\frac{D_r(mz, q/m)}{m^r} = \sum_{j=0}^{m-1} e^{jq/m} D_r(z + j/m, q), \tag{3.9}$$

$$\frac{D_r(0, q/2)}{2^r} = D_r(0, q) + e^{q/2} D_r(\tfrac{1}{2}, q), \tag{3.10}$$

$$\frac{d^\nu}{dq^\nu} e^{zq} D_r(z, q) = e^{zq} D_{r+\nu}(z, q). \tag{3.11}$$

† It is proved later in Section 4 that

$$\lim_{q \to 0} D_r(z, q) = \frac{B_{r+1}(z)}{r+1}.$$

The relations (3.7) and (3.8) show that both $e^{q/2} D_r(\tfrac{1}{2}, q)$ and $D_r(0, q)$ are odd functions of q when r is even and even functions of q when r is odd, except for $D_0(0, q)$ in which case $D_0(0, q) + \tfrac{1}{2}$ is odd in q.

The equations (1.3) and (1.4), however, do not hold. Instead, writing $y + z$ for z in (3.2), we obtain

$$\sum_{r=0}^{\infty} \frac{t^r}{r!} D_r(y + z) = \frac{e^{(y+z)t}}{e^q e^t - 1} - \frac{e^{-(y+z)q}}{q + t}$$

$$= e^{zt}\left\{ \frac{e^{yt}}{e^q e^t - 1} - \frac{e^{-yq}}{q + t} \right\} + \frac{e^{-yq}}{q + t}(e^{zt} - e^{-zq}). \tag{3.12}$$

But, expanding in powers of t, it may be shown that

$$e^{zt}\left\{ \frac{e^{yt}}{e^q e^t - 1} - \frac{e^{-yq}}{q + t} \right\} = \sum_{r=0}^{\infty} \frac{t^r}{r!}\left\{ \sum_{j=0}^{r} \binom{r}{j} D_j(y)\, z^{r-j} \right\}$$

$$= \sum_{r=0}^{\infty} \frac{t^r}{r!} \{D(y) + z\}^r, \tag{3.13}$$

symbolically; and also that

$$\frac{e^{-yq}}{q + t}(e^{zt} - e^{-zq}) = e^{-yq} \sum_{r=0}^{\infty} \frac{t^r}{r!}\left[r!\left(\frac{-1}{q}\right)^{r+1}\left\{ e^{-zq} - \sum_{j=0}^{r} \frac{(-zq)^j}{j!} \right\} \right]$$

$$= \sum_{r=0}^{\infty} \frac{t^r}{r!}\left\{ r!\, e^{-yq}\, z^{r+1} \sum_{j=0}^{\infty} \frac{(-zq)^j}{(r+1+j)!} \right\}. \tag{3.14}$$

Hence, using (3.13) and (3.14) in (3.12), and equating the coefficients of t^r in the resulting equation, we obtain the somewhat different formula

$$D_r(y + z) = \{D(y) + z\}^r + r!\, e^{-yq}\, z^{r+1} \sum_{j=0}^{\infty} \frac{(-zq)^j}{(r+1+j)!}, \tag{3.15}$$

which, for $y = 0$, yields

$$D_r(z) = \{D(0) + z\}^r + r!\, z^{r+1} \sum_{j=0}^{\infty} \frac{(-zq)^j}{(r+1+j)!}. \tag{3.16}$$

4. *Formulae for the calculation of* $C_r(z, q)$ *and* $D_r(z, q)$

In the following we display some formulae for the computation of $C_r(z, q)$; the corresponding value of $D_r(z, q)$ is then given by (3.1),

$$D_r(z, q) = C_r(z, q) + r!\, (-1/q)^{r+1} e^{-zq}.$$

An asymptotic formula for $D_r(0, q)$ is given in Section 6.

The formulae for $C_r(0, q)$ and $D_r(0, q)$ may also be used to compute the coefficients in the formulae (1.4) and (3.16) for $C_r(z, q)$ and $D_r(z, q)$. For $z = \tfrac{1}{2}$, use may instead be made of the simpler formulae (1.13) and (3.10).

The equation (1.9) may be written in the symbolic form as

$$\{C(0)+1\}^r - e^{-q}C_r(0) = 0 \quad (r>0), \tag{4.1}$$

from which we may obtain, step by step, the numbers $C_r(0)$, starting initially with

$$C_0(0) = \frac{1}{e^q-1}. \tag{4.2}$$

Again, since (Jordan, 1939, p. 32)

$$\frac{\mathrm{d}^r}{\mathrm{d}t^r}u(e^t)\bigg|_{t=0} = \sum_{j=0}^{r}\mathscr{S}_r^j u^{(j)}(1),$$

where $\mathscr{S}_r^j$ denotes a Stirling number of the second kind, we have

$$C_r(0) = e^{-q}\frac{\mathrm{d}^r}{\mathrm{d}t^r}\left\{\frac{1}{e^t-e^{-q}}\right\}\bigg|_{t=0} \tag{4.3}$$

$$= -e^{-q}\sum_{j=0}^{r}\frac{j!\,\mathscr{S}_r^j}{(e^{-q}-1)^{j+1}}; \tag{4.4}$$

or, alternatively,

$$\left.\begin{aligned}
C_r(0) &= (-)^r e^q \sum_{j=0}^{r}\frac{j!\,\mathscr{S}_r^j}{(e^q-1)^{j+1}} \quad (r\geqslant 1),\\[2mm]
C_0(0) &= \frac{1}{e^q-1}
\end{aligned}\right\} \tag{4.5}$$

by virtue of (1.11).

Using (4.5) the first six C-numbers may be enumerated as follows:

$$\left.\begin{aligned}
C_0(0) &= \frac{1}{e^q-1},\\[2mm]
C_1(0) &= -\frac{e^q}{(e^q-1)^2},\\[2mm]
C_2(0) &= \frac{e^q}{e^q-1}\left\{\frac{1}{e^q-1}+\frac{2}{(e^q-1)^2}\right\},\\[2mm]
C_3(0) &= -\frac{e^q}{e^q-1}\left\{\frac{1}{e^q-1}+\frac{6}{(e^q-1)^2}+\frac{6}{(e^q-1)^3}\right\},\\[2mm]
C_4(0) &= \frac{e^q}{e^q-1}\left\{\frac{1}{e^q-1}+\frac{14}{(e^q-1)^2}+\frac{36}{(e^q-1)^3}+\frac{24}{(e^q-1)^4}\right\},\\[2mm]
C_5(0) &= -\frac{e^q}{e^q-1}\left\{\frac{1}{e^q-1}+\frac{30}{(e^q-1)^2}+\frac{150}{(e^q-1)^3}+\frac{240}{(e^q-1)^4}+\frac{120}{(e^q-1)^5}\right\}.
\end{aligned}\right\} \tag{4.6}$$

For $\mathrm{Re}\{q\} > 0$, we have from (1.18)

$$C_r(z, q) = \mathrm{e}^{-zq} \frac{\mathrm{d}^r}{\mathrm{d}q^r} \left\{ \frac{\mathrm{e}^{zq}}{\mathrm{e}^q - 1} \right\} \tag{4.7}$$

$$= (-)^r \sum_{m=1}^{\infty} (m - z)^r \mathrm{e}^{-mq}, \tag{4.8}$$

a formula which is useful if $\mathrm{Re}\{q\}$ is not too near zero.

Again, since by (1.11)

$$C_r(z, -q) = (-)^{r+1} C_r(-z, q) - z^r,$$

this same formula (4.8) may also be used for the cases where $\mathrm{Re}\{q\} < 0$.

Alternative formulae for $C_r(z)$ and $D_r(z)$, which are useful when $|q|$ is small, are easily derived thus:

$$C_r(z, q) = \mathrm{e}^{-zq} \frac{\mathrm{d}^r}{\mathrm{d}q^r} \left\{ \frac{1}{q} + \sum_{j=0}^{\infty} \frac{B_{j+1}(z)}{(j+1)!} q^j \right\} \quad \text{(from (4.7))}$$

$$= \mathrm{e}^{-zq} \left\{ -r! \left(\frac{-1}{q} \right)^{r+1} + \sum_{j=0}^{\infty} \frac{B_{j+r+1}(z)}{(j+r+1)j!} q^j \right\} \quad (\text{convgt.} |q| < 2\pi). \tag{4.9}$$

Hence also, from (3.1),

$$D_r(z, q) = \mathrm{e}^{-zq} \sum_{j=0}^{\infty} \frac{B_{j+r+1}(z)}{(j+r+1)j!} q^j \quad (\text{convgt.} |q| < 2\pi), \tag{4.10}$$

which also yields, on taking $q = 0$, the important relationship

$$D_r(z, 0) = \frac{B_{r+1}(z)}{r+1}, \tag{4.11}$$

as mentioned previously in Section 3.

We may notice that the formulae (4.9 and 10) assume simpler forms for $z = 0$ and $z = \frac{1}{2}$. Thus (see also remarks at the beginning of this section) the computation of $C_r(z, q)$ and $D_r(z, q)$ is, in general, simpler when $z = 0$ or $\frac{1}{2}$, an exception being the formula (4.8).

When q is of the form $q = 2\pi i \nu / m$ (ν, m integral, ν/m not an integer), we may also make use of the formula (1.14),

$$C_r(z, 2\pi i \nu / m) = \frac{m^r}{r+1} \sum_{j=0}^{m-1} \mathrm{e}^{2\pi i \nu j / m} B_{r+1} \left(\frac{z+j}{m} \right),$$

particularly when we require $C_r(z, 2\pi i \nu / m)$ for a set of integer values of ν, since in that case we shall require only one set of values of $B_{r+1}\{(z+j)/m\}$, for $j = 0, 1, \ldots, m-1$, for their computation. The formula is therefore convenient for obtaining the coefficients $L_r(s, i\nu)$ given by equation (3.10) of

Chapter I when, for example, the formulae in Section 4 of Chapter I are being used to calculate the successive Fourier coefficients (cf. Chapter I, Section 5).

In concluding the present section we may note that by virtue respectively of (1.15) and (3.11) the Taylor series for $C_r(z, q+\varepsilon)$ and $D_r(z, q+\varepsilon)$ assume the forms

$$C_r(z, q+\varepsilon) = \mathrm{e}^{-z\varepsilon} \sum_{\nu=0}^{\infty} \frac{\varepsilon^\nu}{\nu!} C_{r+\nu}(z, q) \tag{4.12}$$

and

$$D_r(z, q+\varepsilon) = \mathrm{e}^{-z\varepsilon} \sum_{\nu=0}^{\infty} \frac{\varepsilon^\nu}{\nu!} D_{r+\nu}(z, q), \tag{4.13}$$

and that for the computation of $C_r(z, q+\mathrm{i}\pi)$ it is convenient to replace (4.9) by

$$C_r(z, q+\mathrm{i}\pi) = -\mathrm{e}^{-zq} \frac{\partial^r}{\partial q^r} \left(\frac{\mathrm{e}^{zq}}{\mathrm{e}^q+1} \right)$$

$$= -\frac{\mathrm{e}^{-zq}}{2} \frac{\partial^r}{\partial q^r} \left\{ \sum_{j=0}^{\infty} \frac{E_j(z)}{j!} q^j \right\}$$

$$= -\frac{\mathrm{e}^{-zq}}{2} \sum_{j=0}^{\infty} \frac{E_{j+r}(z)}{j!} q^j. \tag{4.14}$$

Interpolatable tables of $D_0(0, \pm q) + \frac{1}{2} = \lambda_0(\pm q)$ and $D_r(0, \pm q) = \lambda_r(\pm q)$ $(1 \leqslant r \leqslant 11)$ for $q = u$ and $q = iv$ are given for u, $v = 0\ (0.02)\ 1$ in Appendix C.

5. *Upper bounds for* $|\mathrm{e}^{sq} D_r(s, q)|$ *and* $|\mathrm{e}^{sq} C_r(s, q)|$ $(0 \leqslant s \leqslant 1, |v| < 2\pi, r \geqslant 1)$

We have (cf. Chapter I, eqns (3.2–10) and Appendix A, eqns (A1) and (A2)), for $0 \leqslant s \leqslant 1,\ r > 0$ or $0 < s < 1,\ r \geqslant 0$,

$$\mathrm{e}^{sq} D_r(s, q) = \frac{r!}{(-2\pi\mathrm{i})^{r+1}} \{(-)^r \bar{J}_r(s) - J_r(s)\}, \tag{5.1}$$

where

$$J_r(s) = \mathrm{e}^{-2\pi\mathrm{i}s} \Phi(\mathrm{e}^{-2\pi\mathrm{i}s}, r+1, 1-\mathrm{i}q/2\pi) \tag{5.2}$$

$$= \sum_{m=1}^{\infty} \frac{\mathrm{e}^{-2\pi\mathrm{i}sm}}{(m - \mathrm{i}q/2\pi)^{r+1}}. \tag{5.3}$$

Hence, for $q = u + iv$ and $r \geqslant 1$,

$$|J_r(s)| \leqslant \sum_{m=1}^{\infty} \frac{1}{\{(m + v/2\pi)^2 + (u/2\pi)^2\}^{(r+1)/2}} \tag{5.4}$$

which yields, for $|v| < 2\pi$,

$$|J_r(s)| \leqslant \sum_{m=1}^{\infty} \frac{1}{(m + v/2\pi)^{r+1}} = \zeta(r+1, 1+v/2\pi), \tag{5.5}$$

which is more conservative.

A better upper bound for $|J_r(s)|$ may be derived by firstly noticing that for $|v| < 2\pi$

$$\sum_{m=1}^{\infty} \frac{1}{\{(m+v/2\pi)^2 + (u/2\pi)^2\}^{(r+1)/2}}$$

$$\leqslant \frac{1}{\{(1+v/2\pi)^2 + (u/2\pi)^2\}^{(r+1)/2}} + \int_1^{\infty} \frac{dt}{\{(t+v/2\pi)^2 + (u/2\pi)^2\}^{(r+1)/2}}, \tag{5.6}$$

since the integrand is a monotonically decreasing function.

But for $k > 0$, $y > \frac{1}{2}$,

$$\int_k^{\infty} \frac{dt}{(t^2+a^2)^y} = \frac{1}{2} \int_{k^2+a^2}^{\infty} \frac{dt}{t^y(t-a^2)^{\frac{1}{2}}}$$

$$\leqslant \frac{(k^2+a^2)^{\frac{1}{2}}}{2k} \int_{k^2+a^2}^{\infty} \frac{dt}{t^{y+\frac{1}{2}}} = \frac{1}{k(2y-1)(k^2+a^2)^{y-1}}. \tag{5.7}$$

Hence we obtain, using (5.7) in (5.6) and from (5.4),

where
$$\left.\begin{aligned}
|J_r(s)| &\leqslant \{\chi(q)\}^{r-1}\left[\{\chi(q)\}^2 + \frac{1}{r(1+v/2\pi)}\right], \\[2mm]
\chi(q) &= \left|\frac{1}{1-iq/2\pi}\right| = \frac{1}{\{(1+v/2\pi)^2 + (u/2\pi)^2\}^{\frac{1}{2}}}.
\end{aligned}\right\} \tag{5.8}$$

Corresponding inequalities for $|\bar{J}_r(s)|$ are obtained by replacing q by $-q$ in the inequalities (5.4, 5 and 8) for $|J_r(s)|$.

Hence there follow from (5.1), for $0 \leqslant s \leqslant 1$, $r > 0$, $|v| < 2\pi$,

$$|e^{sq} D_r(s,q)| \leqslant \frac{r!}{(2\pi)^{r+1}} \{\zeta(r+1, 1-v/2\pi) + \zeta(r+1, 1+v/2\pi)\}, \tag{5.9}$$

$$|e^{sq} D_r(s,q)| \leqslant \frac{r!}{(2\pi)^{r+1}} \left(\{\chi(-q)\}^{r-1}\left[\{\chi(-q)\}^2 + \frac{1}{r(1-v/2\pi)}\right]\right.$$

$$\left. + \{\chi(q)\}^{r-1}\left[\{\chi(q)\}^2 + \frac{1}{r(1+v/2\pi)}\right]\right), \tag{5.10}$$

where $\chi(q)$ is given by (5.8).

It is useful to notice that if q is purely real, i.e. $q = u$, the inequalities (5.9 and 10) assume the simpler forms

$$|\,\mathrm{e}^{su}\,D_r(s,u)\,| \leqslant \frac{r!}{2^r}\,\frac{\zeta(r-1)}{\pi^{r+1}} = \left|\frac{B_{r+1}}{r+1}\right| \quad \text{(if } r \text{ odd)}, \tag{5.11}$$

and

$$|\,\mathrm{e}^{su}\,D_r(s,u)\,| \leqslant \frac{r!}{2^r\,\pi^{r+1}}\,\{\chi(u)\}^{r-1}\left[\{\chi(u)\}^2 + \frac{1}{r}\right], \tag{5.12}$$

where, from (5.8),

$$\chi(u) = \frac{1}{\{1 + (u/2\pi)^2\}^{\frac{1}{2}}}. \tag{5.13}$$

On the other hand if q is purely imaginary, i.e. $q = iv$, there follows from (5.1 and 3),

$$D_r(0,iv) = \frac{r!}{(-2\pi i)^{r+1}}\{(-)^r\,\zeta(r+1,1-v/2\pi) - \zeta(r+1,1+v/2\pi)\}, \tag{5.14}$$

the comparison of which with (5.9) shows that for r odd

$$|\,D_r(s,iv)\,| \leqslant |\,D_r(0,iv)\,|, \tag{5.15}$$

while (5.10) may be reduced to the simpler but more conservative upper bound

$$|\,D_r(s,iv)\,| < \frac{4r!}{(2\pi - |v|)^{r+1}}. \tag{5.16}$$

From (3.1) there follows

$$|\,\mathrm{e}^{sq}\,C_r(s,q)\,| \leqslant |\,\mathrm{e}^{sq}\,D_r(s,q)\,| + \frac{r!}{|q|^{r+1}}, \tag{5.17}$$

which provides a set of upper bounds for $|\,\mathrm{e}^{sq}\,C_r(s,q)\,|$ corresponding to the upper bounds for $|\,\mathrm{e}^{sq}\,D_r(s,q)\,|$ obtained in this section.

For example, the use of (5.9) in (5.17) yields

$$|\,\mathrm{e}^{sq}\,C_r(s,q)\,| \leqslant \frac{r!}{(2\pi)^{r+1}}\{\zeta(r+1,1-|v|/2\pi) + \zeta(r+1,|v|/2\pi)\}. \tag{5.18}$$

If q is purely imaginary, i.e. $q = iv$ we have from (5.14) and (3.1)

$$C_r(0,iv) = \frac{r!}{(-2\pi i)^{r+1}}\{(-)^r\,\zeta(r+1,1-v/2\pi) - \zeta(r+1,v/2\pi)\}, \tag{5.19}$$

the comparison of which with (5.18) shows that for $0 < |v| < 2\pi$ and r odd

$$|\,C_r(s, \pm iv)\,| \leqslant |\,C_r(0,iv)\,|. \tag{5.20}$$

When q is purely real, i.e. $q = u$, we have from (4.8) that for $u > 0$

$$C_r(s,u) = (-)^r\,|\,C_r(s,u)\,|, \tag{5.21}$$

where $|C_r(s,u)|$ is monotonic decreasing in $-\infty < s \leqslant 1$, and that for $u < 0$

$$C_r(s,u) = -|C_r(s,u)|, \tag{5.22}$$

where $|C_r(s,u)|$ is monotonic increasing in $0 \leqslant s < \infty$.

For $q = u + i\pi$, u real, (5.18) reduces to

$$|e^{su} C_r(s, u+i\pi)| \leqslant \frac{r!}{2^r \pi^{r+1}} (2^{r+1} - 1)\, \zeta(r+1), \tag{5.23}$$

while the use of (5.10) in (5.17) yields

$$|e^{su} C_r(s, u+i\pi)| \leqslant 3r!/\pi^{r+1}. \tag{5.24}$$

In concluding this section it may be noted that by virtue of (3.7) and (1.10) any upper bounds for $|e^{sq} D_r(s,q)|$ and $|e^{sq} C_r(s,q)|$ in $0 \leqslant s \leqslant 1$ must also be the respective upper bounds for $|e^{-sq} D_r(s,-q)|$ and $|e^{-sq} C_r(s,-q)|$ in the same interval $0 \leqslant s \leqslant 1$.

6. *An asymptotic formula for* $D_r(0,q)$

If we apply the Euler–Maclaurin formula to the integral

$$\int_k^\infty \frac{dt}{(t - iq/2\pi)^{r+1}},$$

we obtain for $r \geqslant 1$, $h = 1$,

$$\sum_{m=k}^\infty \frac{1}{(m - iq/2\pi)^{r+1}} = \frac{1}{r(k - iq/2\pi)^r} + \frac{1}{2(k - iq/2\pi)^{r+1}}$$

$$+ \sum_{\nu=1}^{n-1} \frac{B_{2\nu}}{(2\nu)!} \frac{(r+1)(r+2)\ldots(r+2\nu-1)}{(k - iq/2\pi)^{r+2\nu}} + R, \tag{6.1}$$

where (Steffensen, 1950, p. 135)

$$R = -\frac{(r+1)(r+2)\ldots(r+2n)}{(2n)!} \int_0^\infty \frac{\bar{B}_{2n}(t) - B_{2n}}{(k+t - iq/2\pi)^{r+2n+1}}\, dt, \tag{6.2}$$

where $\bar{B}_{2n}(t)$ denotes the periodic Bernoulli function.

Since $|\bar{B}_{2n}(t) - B_{2n}| \leqslant (2 - 2^{1-2n})|B_{2n}|$, there follows, for $q = u + iv$,

$$\left| \int_0^\infty \frac{\bar{B}_{2n}(t) - B_{2n}}{(k+t - iq/2\pi)^{r+2n+1}}\, dt \right|$$

$$\leqslant (2 - 2^{1-2n})|B_{2n}| \int_0^\infty \frac{dt}{\{(t+k+v/2\pi)^2 + (u/2\pi)^2\}^{(r+2n+1)/2}}$$

$$\leqslant \frac{(2 - 2^{1-2n})|B_{2n}|}{(k+v/2\pi)(r+2n)\{(k+v/2\pi)^2 + (u/2\pi)^2\}^{(r+2n-1)/2}}, \tag{6.3}$$

from (5.7).

Hence using (6.3) in (6.2) we obtain

$$|R| \leqslant T_n \rho_n,$$ (6.4)

where, writing

$$\alpha(q, k) = \frac{1}{|k - iq/2\pi|},$$ (6.5)

we have that

$$T_n = \frac{|B_{2n}| \, (r+1) \, (r+2) \dots (r+2n-1)}{(2n)!} \{\alpha(q, k)\}^{r+2n}$$ (6.6)

is the modulus of the first term neglected in the correction series in (6.1), and

$$\rho_n = \frac{(2 - 2^{1-2n})}{(k + v/2\pi) \, \alpha(q, k)}.$$ (6.7)

Hence, using (6.1) in (5.3), we obtain the following asymptotic formula for $J_r(0)$ for $r \geqslant 1$:

$$J_r(0) \sim \sum_{m=1}^{k-1} \frac{1}{(m - iq/2\pi)^{r+1}} + \frac{1}{2} \frac{1}{(k - iq/2\pi)^{r+1}}$$

$$+ \frac{1}{r(k - iq/2\pi)^r} + \sum_{v=1}^{\infty} \frac{B_{2v}}{(2v)!} \frac{(r+1)\,(r+2)\dots(r+2v-1)}{(k - iq/2\pi)^{r+2v}},$$ (6.8)

where the truncation of the second series by neglecting all terms for which $v \geqslant n$ will give rise to the error R as given by (6.2). A relationship between the upper bound for $|R|$ and the modulus of the first neglected term is given by (6.4).

The corresponding formula for $\bar{J}_r(0)$ may be obtained by writing $-q$ for q in (6.8), so that, in conjunction with (5.1), this formula (6.8) is, in fact, an asymptotic formula for $D_r(0, q)$ for $r \geqslant 1$.

7. *The general rectangular formula with the exponential weighting function*

Let $\bar{D}_r(z)$, a damped periodic function with unit period, be defined for real z by

$$\begin{aligned} \bar{D}_r(z) &= D_r(z) \quad (0 \leqslant z < 1), \\ \bar{D}_r(z+1) &= e^{-q} \bar{D}_r(z) \quad (\text{for all } z). \end{aligned} \Bigg\}$$ (7.1)

Since, from (3.6),

$$D_r(1) = e^{-q} D_r(0) \quad (r > 0),$$ (7.2)

it therefore follows that $\bar{D}_r(z)$ is a continuous function of z, for all z, provided that $r > 0$. $\bar{D}_0(z)$ is, however, discontinuous at $z = 0, \pm 1, \pm 2, \dots$

Again we may derive from (3.3) that

$$\frac{\mathrm{d}}{\mathrm{d}z}\,\overline{D}_r(z) = r\overline{D}_{r-1}(z),\tag{7.3}$$

so that $\mathrm{d}/\mathrm{d}z\,\overline{D}_r(z)$ is continuous for all z, so long as $r > 1$. But

$$\frac{\mathrm{d}}{\mathrm{d}z}\,\overline{D}_1(z) = \overline{D}_0(z)$$

is discontinuous at $z = 0,\, \pm 1,\, \pm 2,\, \ldots$.

Let us now consider the expression

$$R_m = -\frac{h^{m+2}\,\mathrm{e}^{(sq+px)}}{m!}\int_0^1 \overline{D}_m(s-t)f^{(m+1)}(x+ht)\,\mathrm{d}t.\tag{7.4}$$

Integrating by parts and using (7.1) and (7.3) we obtain, for $m > 0$ and $q = ph$,

$$R_m = -\frac{h^{m+1}\,\mathrm{e}^{sq}\,\overline{D}_m(s)}{m!}\{\mathrm{e}^{p(x+h)}f^{(m)}(x+h) - \mathrm{e}^{px}f^{(m)}(x)\} + R_{m-1}.\tag{7.5}$$

Hence, treating R_{m-1} in the same way and so on, we obtain

$$R_m = -\sum_{r=m}^1 \frac{h^{r+1}\,\mathrm{e}^{sq}\,\overline{D}_r(s)}{r!}\{\mathrm{e}^{p(x+h)}f^{(r)}(x+h) - \mathrm{e}^{px}f^{(r)}(x)\} + R_0.\tag{7.6}$$

Since $\overline{D}_0(s-t)$ is discontinuous at $t = s$, we now assume that

$$0 \leqslant s \leqslant 1,\tag{7.7}$$

and use (7.1) to write R_0 in the form

$$R_0 = -h^2\,\mathrm{e}^{(sq+px)}\left\{\int_0^s D_0(s-t)f'(x+ht)\,\mathrm{d}t + \int_s^1 \mathrm{e}^q\,D_0(s+1-t)f'(x+ht)\,\mathrm{d}t\right\}$$

$$= -h^2\,\mathrm{e}^{(sq+px)}\left\{\int_0^s \left(C_0(0) - \frac{\mathrm{e}^{-q(s-t)}}{q}\right)f'(x+ht)\,\mathrm{d}t\right.$$

$$\left. + \int_s^1 \mathrm{e}^q\left(C_0(0) - \frac{\mathrm{e}^{-q(s+1-t)}}{q}\right)f'(x+ht)\,\mathrm{d}t\right\},$$

from (3.1), which yields, on integration by parts,

$$R_0 = h\,\mathrm{e}^{p(x+sh)}f(x+sh) - h\,\mathrm{e}^{sq}\,D_0(s)\{\mathrm{e}^{p(x+h)}f(x+h) - \mathrm{e}^{px}f(x)\}$$

$$-h\int_0^1 \mathrm{e}^{p(x+ht)}f(x+ht)\,\mathrm{d}t.$$

Inserting this expression for R_0 in (7.6) and writing t for $x+ht$ in the integral, we obtain the formula

$$h\,e^{p(x+sh)}f(x+sh) - \int_x^{x+h} e^{pt}f(t)\,dt$$

$$= \sum_{r=0}^{m} \frac{h^{r+1}\,e^{sq}\,D_r(s)}{r!}\{e^{p(x+h)}f^{(r)}(x+h) - e^{px}f^{(r)}(x)\} + R_m. \quad (7.8)$$

If in (7.8) we write $x_r = x_0 + rh$ for x and sum from $r = 0$ to $r = n-1$, we rederive the *general rectangular formula* (3.7) *of Chapter I for the exponential weighting function* in the form

$$h\sum_{r=0}^{n-1} e^{p(x_r+sh)}f(x_r+sh) - \int_{x_0}^{x_n} e^{pt}f(t)\,dt$$

$$= \sum_{r=0}^{m} \frac{h^{r+1}\,e^{sq}\,D_r(s)}{r!}\{e^{px_n}f^{(r)}(x_n) - e^{px_0}f^{(r)}(x_0)\} + R(s,q), \quad (7.9)$$

where now the remainder $R(s,q)$ has the different expression

$$R(s,q) = -\sum_{r=0}^{n-1} \frac{h^{m+2}\,e^{sq}\,e^{px_r}}{m!} \int_0^1 \overline{D}_m(s-t)f^{(m+1)}(x_r+ht)\,dt \quad (7.10)$$

$$= -\frac{h^{m+2}\,e^{sq}\,e^{px_0}}{m!} \int_0^n \overline{D}_m(s-t)f^{(m+1)}(x_0+ht)\,dt. \quad (7.11)$$

It should be noted here that, whereas the form of the remainder $R(s,q)$ given by the complex variable method of Chapter I involves values of $f(t)$ and its derivatives in the complex plane, the alternative form (7.10–11) obtained above requires for its application information about $f^{(m+1)}(t)$ for $x_0 \leqslant t \leqslant x_n$ only.

This latter form holds only if the same number of correction terms are retained at both the end points x_0 and x_n, i.e. when, in the notation of Sections 3 and 4 of Chapter I, $m_0 = m_n = m$; and, subject to this restriction, the form (7.10–11) for $R(s,q)$ is valid for all the formulae of Sections 3 and 4 of Chapter I, involving the exponential and allied weighting functions.

8. *Error bounds for the formulae of Sections 3 and 4 of Chapter I*

Since

$$e^{sq}\int_0^1 \overline{D}_m(s-t)f^{(m+1)}(x_r+ht)\,dt$$

$$= \int_0^s e^{q(s-t)}D_m(s-t)\,e^{qt}f^{(m+1)}(x_r+ht)\,dt$$

$$+ \int_s^1 e^{q(s+1-t)}D_m(s+1-t)\,e^{qt}f^{(m+1)}(x_r+ht)\,dt$$

(see (7.1)), there follows

$$\left| e^{sq} \int_0^1 \overline{D}_m(s-t) f^{(m+1)}(x_r + ht)\, \mathrm{d}t \right| \leqslant M \mu_r(q),$$

where

$$M = \max | e^{sq} D_m(s,q) | \quad (0 \leqslant s \leqslant 1), \tag{8.1}$$

and

$$\mu_r(q) = \max | e^{qt} f^{(m+1)}(x_r + ht) | \quad (0 \leqslant t \leqslant 1). \tag{8.2}$$

Hence for $p = \alpha + i\beta$, $q = u + iv$ (7.10) yields

$$| R(s,q) | \leqslant \frac{h^{m+2}}{m!} M \sum_{r=0}^{n-1} e^{\alpha x_r} \mu_r(q), \tag{8.3}$$

$$\leqslant \begin{cases} \dfrac{h^{m+2}}{m!} M \mu^*(q)\, e^{\alpha x_0}(1 - e^{un})/(1 - e^u) \quad (\alpha \neq 0) & \tag{8.4} \\[3mm] \dfrac{h^{m+2}}{m!} M \mu^*(q)\, n \quad (\alpha = 0), & \tag{8.5} \end{cases}$$

where

$$\mu^*(q) = \max \{ \mu_r(q) \} \quad (0 \leqslant r \leqslant n-1). \tag{8.6}$$

Some upper bounds for M, we may note, are given in Section 5.

If $f^{(m+1)}(t)$ retains its sign in $x_0 \leqslant t \leqslant x_n$ and if $p = i\beta$ $(q = iv)$, we may obtain from (7.11)

$$| R(s, iv) | \leqslant \frac{h^{m+1}}{m!} M | f^{(m)}(x_n) - f^{(m)}(x_0) |. \tag{8.7}$$

It should be noticed that these upper bounds (8.3, 4, 5 and 7) for the remainder term in (7.9) are independent of s and therefore apply equally well to the remainder term in each of the formulae (3.7, 12 and 14) of Chapter I, with $m_n = m_0 = m$, for the exponential weighting function.

Similar upper bounds for the remainder terms in the formulae of Section 4 of Chapter I may also be obtained:

Since M given by (8.1) is unaltered if q is replaced by $-q$ (see (3.7)), there follows from (8.3 and 4)

$$\tfrac{1}{2} | R(s, u) \pm R(s, -u) | \leqslant \tfrac{1}{2} \{ | R(s, u) | + | R(s, -u) | \}$$

$$\leqslant \frac{h^{m+2}}{m!} M \sum_{r=0}^{n-1} \cosh (\alpha x_r)\, \mu_r(| u |) \tag{8.8}$$

$$\leqslant \frac{h^{m+2}}{m!} M \mu^*(| u |) \frac{\sinh (\alpha x_n - u/2) - \sinh (\alpha x_0 - u/2)}{2 \sinh (u/2)} \tag{8.9}$$

for the remainder terms in the formulae (4.2, 4 and 7) of Chapter I, with $m_n = m_0 = m$, for the weighting functions $\cosh(\alpha x)$ and $\sinh(\alpha x)$.

Similarly from (8.3 and 5) we obtain

$$\tfrac{1}{2}|R(s, iv) \pm R(s, -iv)| \leqslant \tfrac{1}{2}\{|R(s, iv)| + |R(s, -iv)|\}$$

$$\leqslant \frac{h^{m+2}}{m!} M \sum_{r=0}^{n-1} \mu_r \tag{8.10}$$

$$\leqslant \frac{h^{m+2}}{m!} M \mu^* n, \tag{8.11}$$

where

$$\mu_r = \max |f^{(m+1)}(x_r + ht)| \quad (0 \leqslant t \leqslant 1)$$

and

$$\mu^* = \max\{\mu_r\} \quad (0 \leqslant r \leqslant n-1),$$

for the remainder terms in the formulae (4.2, 4 and 7) of Chapter I, with $m_n = m_0 = m$, for the weighting functions $\cos(\beta x)$ and $\sin(\beta x)$.

If $f^{(m+1)}(t)$ keeps its sign in $x_0 \leqslant t \leqslant x_n$, we also have from (8.7)

$$\tfrac{1}{2}|R(s, iv) \pm R(s, -iv)| \leqslant \frac{h^{m+1}}{m!} M |f^{(m)}(x_n) - f^{(m)}(x_0)|. \tag{8.12}$$

It should be noted here that both the forms (8.5) and (8.11) will become indeterminate whenever the interval of integration is infinite.

9. Error estimates for the formulae with the exponential and allied weighting functions, comparison with other formulae, and numerical examples

In this section we consider the remainder term

$$R(s, q) = -\frac{h^{n+2} e^{px_0}}{m!} \int_0^1 e^{sq} \bar{D}_m(s-t) f^{(m+1)}(x_0 + ht)\, dt \tag{9.1}$$

in the general rectangular formula (7.9), with $n = 1$, and compare it with the remainder terms

$$E_1 = -\frac{h^{m+2}}{(m+1)!} \int_0^1 \bar{B}_{m+1}(s-t) F^{(m+1)}(x_0 + ht)\, dt, \tag{9.2}$$

where

$$F(x) = e^{px} f(x), \tag{9.3}$$

obtained on using the Euler–Maclaurin formula (cf. Steffensen, 1950;

Nörlund, 1924), and

$$E_2 = \frac{h^{m+2}\,\mathrm{e}^{px_0}}{(-q)^{m+1}} \int_0^1 \mathrm{e}^{qt} f^{(m+1)}(x_0 + ht)\,\mathrm{d}t, \tag{9.4}$$

obtained on using the method of successive integration by parts (see Section 10), for the evaluation of the same integral

$$\int_{x_0}^{x_0+h} \mathrm{e}^{px} f(x)\,\mathrm{d}x,$$

the remainder term in each case representing the error committed when all derivatives beyond the mth are neglected.

The conclusions, which are relevant to all the formulae of Sections 3 and 4 of Chapter I, are then exemplified with the help of simple numerical examples.

We have, from (8.3), the estimate

$$|R(s,q)| = O[h^{m+2}\,M/m!]. \tag{9.5}$$

Again, from (9.3),

$$F^{(m+1)}(x+ht) = p^{m+1} \sum_{j=0}^{m+1} \binom{m+1}{j} p^{-j}\,\mathrm{e}^{p(x+ht)} f^{(j)}(x+ht),$$

so that

$$|F^{(m+1)}(x+ht)| \begin{cases} = O\{|p^{m+1}|\} & (\text{if } |p| \text{ large}) \\ = O\{1\} & (\text{if } |p| \text{ small}), \end{cases} \tag{9.6}$$

and consequently there follows from (9.2)

$$|E_1| \begin{cases} = O\{h\,|q|^{m+1}\,\zeta(m+1)/2^m\,\pi^{m+1}\} & (\text{if } |p| \text{ large}) \\ = O\{h^{m+2}\,\zeta(m+1)/2^m\,\pi^{m+1}\} & (\text{if } |p| \text{ small}). \end{cases} \tag{9.7}$$

Similarly, (9.4) leads to

$$|E_2| = O[h^{m+2}/|q|^{m+1}]. \tag{9.8}$$

From (9.5, 7 and 8) we may obtain corresponding estimates for the comparative efficiency factors $|R(s,q)/E_1|$ and $|R(s,q)/E_2|$. These, together with the estimate (9.5) for $|R(s,q)|$, are displayed below in table (9.9) for the special cases $p = \alpha$ and $p = i\beta$, where we have replaced M in (9.5) by its upper bounds (5.11 and 16).

(a) The estimates in (9.9) indicate that *for $p = \alpha$ real*:

(i) The accuracy of the approximations obtained by using the formula (7.9) is unlikely to be adversely affected by the largeness of $|\alpha|$ (on the contrary, the opposite is suggested by the more realistic upper bound (5.12) for M).

(ii) The formula (7.9) may in general be expected to yield approximations that are better than those obtained by a similarly truncated Euler–Maclaurin formula, the comparative efficiency factor $|R(s,q)/E_1|$ becoming smaller with increase in $|\alpha|$ and the number of derivatives retained.

(iii) As compared, however, with a similarly truncated formula employing successive integration by parts (this method has a fixed upper limit to the accuracy obtainable), the formula (7.9) may be expected to yield better approximations whenever $h<|2\pi/\alpha|$. The comparative efficiency factor $|R(s,q)/E_2|$ decreases, in this case, with reduction in $|u|$ and, for $h<|2\pi/\alpha|$, with an increase in the number of derivatives retained.

Table (9.9)

	(a) $p = \alpha, q = \alpha h = u$	(b) $p = i\beta, q = i\beta h = iv\ (\,|v| < 2\pi)$
$|R(s,q)|$	$O[h^{m+2}\,\zeta(m+1)/2^m\,\pi^{m+1}]$	$O[4h^{m+2}/(2\pi-|v|)^{m+1}]$
$|R(s,q)/E_1|$	$O[\,|1/\alpha|^{m+1}]$ (if $|\alpha|$ large) $O[1]$ (if $|\alpha|$ small)	$O[2/\{(1-|v/2\pi|)|\beta|\}^{m+1}]$ (if $|\beta|$ large) $O[2/(1-|v/2\pi|)^{m+1}]$ (if $|\beta|$ small)
$|R(s,q)/E_2|$	$O[2\,|u/2\pi|^{m+1}]$	$O[4/(\,|2\pi/v|-1)^{m+1}]$

(b) Correspondingly, *for $p = i\beta$ imaginary*, it may be observed from (9.9) that:

(i) On account of the presence of the factor $1/(2\pi-|v|)^{m+1}$ in the estimate for $|R(s,q)|$, the formula (7.9) will not assume its full strength unless $|v/2\pi|$ is significantly less than unity. Thus, in the instances where $|v/2\pi|$ is not already small, a reduction in h will improve the accuracy for two distinct reasons.

(ii) The approximations obtained by the formula (7.9) may be expected to be closer than those obtained by a similarly truncated Euler–Maclaurin formula whenever $|\beta|$ is large enough for the validity of (9.6) and h is

Table (9.11)

(a) $p = \alpha$, $q = \alpha h = u$, $I(\alpha) = \int_0^{\pi} e^{\alpha x} \cos(x)\,dx$

| h | $\alpha = 1$ | | | $\alpha = 4$ | | | $\alpha = 8$ | | |
| | $-10^6\,I(\alpha) = 12070346$ | | | $-10^3\,I(\alpha) = 67471133$ | | | $-10^{-3}\,I(\alpha) = 10120162$ | | |
	Euler–Maclaurin	Formula (3.14) of Chapter I	u	Euler–Maclaurin	Formula (3.14) of Chapter I	u	Euler–Maclaurin	Formula (3.14) of Chapter I	u
$\pi/4$	12070347	12070346	$\pi/4$	67395694	67471132	π	16139840	10120162	2π
$\pi/8$	12070346			67471036	67471133	$\pi/2$	10128468		
$\pi/16$				67471132			10120171		
$\pi/32$				67471133			10120162		
Successive integration by parts	24140693			67471197			10120162		

(b) $p = i\beta$, $q = i\beta h = iv$, $I(\beta) = \int_0^{\pi} \cos(\beta x)\,e^x\,dx$

| h | $\beta = 1$ | | | $\beta = 6$ | | | $\beta = 12$ | | |
| | $-10^6\,I(\beta) = 12070346$ | | | $10^8\,I(\beta) = 59839710$ | | | $10^8\,I(\beta) = 15269443$ | | |
	Euler–Maclaurin	Formula (4.7) of Chapter I	v	Euler–Maclaurin	Formula (4.7) of Chapter I	v	Euler–Maclaurin	Formula (4.7) of Chapter I	v
$\pi/4$	12070347	12070346	$\pi/4$	151257380	61569452	$3\pi/2$	-7397976500	15273418	3π
$\pi/8$	12070346			59892838	59839710	$3\pi/4$	54137277	15271430	$3\pi/2$
$\pi/16$				59839756			15286487	15269443	$3\pi/4$
$\pi/32$				59839710			15269458		
Successive integration by parts	0.0			59839674			15269443		

sufficiently small so that

$$|v/2\pi| < (|\beta| - 1)/|\beta| \quad (\approx 1 \text{ for } |\beta| \text{ large}). \tag{9.10}$$

The comparative efficiency factor $|R(s,q)/E_1|$ diminishes, for any prescribed β, with decrease in h and (provided that (9.10) holds) with an increase in the number of derivatives retained.

(iii) As compared with a similarly truncated formula for successive integration by parts, the formula (7.9) may be expected to yield better approximations whenever $|v| < \pi$. Its comparative efficiency improves with decrease in $|v|$ and, for $h < |\pi/\beta|$, with an increase in the number of derivatives retained.

It may be noted here that, by virtue of (8.8 and 10), the estimates for the comparative efficiency factors and for the remainder term R, associated with the general rectangular formula (4.2) of Chapter I, with $m_0 = m_n = m$ and with the weighting functions $\cosh(\alpha x)$ or $\sinh(\alpha x)$ [and with the weighting functions $\cos(\beta x)$ or $\sin(\beta x)$], are exactly the same as those given, for the special case $q = u$, in column (a) [for the special case $q = iv$, in column (b)] of the table (9.9) for $R(s,q)$.

Thus the remarks following (9.9) provide us with an adequate guide to the effectiveness of all the formulae of Sections 3 and 4 of Chapter I.

In illustration, two integrals have been evaluated:

$$\text{(a)} \quad I(\alpha) = \int_0^{\pi} e^{\alpha x} \cos x \, dx$$

$$\text{(b)} \quad I(\beta) = \int_0^{\pi} \cos(\beta x) \, e^x \, dx$$

for several values of α, β and h. The formulae corresponding to the trapezoidal sum were used and all derivatives beyond the eighth were neglected. The results are given in table (9.11) to eight significant figures.

In the case of $I(\alpha)$, where $q = u$ is real, the contrast in the efficiency of the formulae is sufficiently sharp and requires no further comments.

In the case of $I(\beta)$, where $q = iv$ is imaginary, the formula (4.7) of Chapter I yields the required eight-figure values of $I(\beta)$, for both $\beta = 6$ and $\beta = 12$, when $|v/\pi| = \frac{3}{4}$. As pointed out in the remarks following (9.9), this value ($\frac{3}{4}$) of $|v/\pi|$ is not sufficiently small for the formula to assume its full strength.

Given below are the details for the cases where, for $h = \pi/32$ (so that $|v/\pi| = \frac{3}{8}$ and $\frac{3}{16}$), the efficiency of (4.7) of Chapter I is more in evidence and is fairly independent of the size of β.

4

(i) $\beta = 6$, $h = \pi/32$, $v = 3\pi/16$, $I(\beta) = 0.59839710$

	Euler–Maclaurin	Formula (4.7) of Chapter I	Successive integration by parts
Trapezoidal sum	0.61649001	0.61649001	...
D^0	0	0	0
D^1	−1778319	−1809601	0.61501924
D^2	0	0	0
D^3	−30566	310	−1708387
D^4	0	0	0
D^5	−401	0	47455
D^6	0	0	0
D^7	−4	0	−1318
Σ	0.59839710	0.59839710	0.59839674

(ii) $\beta = 12$, $h = \pi/32$, $v = 3\pi/8$, $I(\beta) = 0.15269443$

	Euler–Maclaurin	Formula (4.7) of Chapter I	Successive integration by parts
Trapezoidal sum	0.17177920	0.17177920	...
D^0	0	0	0
D^1	−1778319	−1908870	0.15375481
D^2	0	0	0
D^3	−123122	393	−106774
D^4	0	0	0
D^5	−6702	0	741
D^6	0	0	0
D^7	−319	0	−5
Σ	0.15269458	0.15269443	0.15269443

10. *Formulae for the summation of series with exponential and allied weighting functions*

The usefulness of the rectangular and trapezoidal formulae for the numerical summation of series—when the integral can be evaluated otherwise—is well known. For sums of the form $\sum e^{rq} f(x_r)$, however, it is possible to obtain formulae the application of which does not involve the evaluation of an integral.

The derivation of these formulae is the same as that of (7.8) with $C_r(z)$ replacing $D_r(z)$. Thus if $\bar{C}_r(z)$, a damped periodic function with unit period, be defined for real z by

$$\left. \begin{aligned} \bar{C}_r(z) &= C_r(z) \quad (0 \leqslant z < 1), \\ \bar{C}_r(z-1) &= \mathrm{e}^{-q} \bar{C}_r(z) \quad \text{(for all } z\text{),} \end{aligned} \right\} \tag{10.1}$$

then, on account of the similarity in the properties of $C_r(z)$ and $D_r(z)$ (see Sections 1 and 3), the function $\bar{C}_r(z)$ retains the properties of $\bar{D}_r(z)$ given in Section 7, and consequently the result of replacing (7.4) by

$$R_m = -\frac{h^{m+2} \mathrm{e}^{(sq+px)}}{m!} \int_0^1 \bar{C}_m(s-t) f^{(m+1)}(x+ht)\, \mathrm{d}t \tag{10.2}$$

is to obtain, in place of (7.8), the formula†

$$h\, \mathrm{e}^{p(x+sh)} f(x+sh) = \sum_{r=0}^{m} \frac{h^{r+1} \mathrm{e}^{sq} C_r(s)}{r!} \{ \mathrm{e}^{p(x+h)} f^{(r)}(x+h) - \mathrm{e}^{px} f^{(r)}(x) \} + R_m, \tag{10.3}$$

where, as before, $q = ph$.

If in (10.3) we write $x_r - sh$ for x and sum from $r = 0$ to $r = n-1$, we obtain on multiplying throughout by e^{-px_0} the desired summation formula in the form

$$\sum_{r=0}^{n-1} \mathrm{e}^{rq} f(x_r) = \sum_{r=0}^{m} \frac{h^r C_r(s,q)}{r!} \{ \mathrm{e}^{nq} f^{(r)}(x_n - sh) - f^{(r)}(x_0 - sh) \} + R(s,q), \tag{10.4}$$

---

† The following formula (10.3) is a generalization of (2.8) to which it reduces when $f(x) \equiv P_m(x)$, a polynomial of degree m in x.

It should also be noted that it is possible to derive (10.3) from (7.8), and vice versa, with the help of the relationship

$$\int_x^{x+h} \mathrm{e}^{pt} f(t)\, \mathrm{d}t = -\sum_{r=0}^{m} h^{r+1}(-1/q)^{r+1} \{ \mathrm{e}^{p(x+h)} f^{(r)}(x+h) - \mathrm{e}^{px} f^{(r)}(x) \}$$

$$+ h^{m+2} \mathrm{e}^{(sq+px)} \int_0^1 (-1/q)^{m+1} \mathrm{e}^{q(t-s)} f^{(m+1)}(x+ht)\, \mathrm{d}t$$

which is obtained by integration by parts.

where

$$R(s,q) = -\frac{h^{m+1}}{m!} \sum_{r=0}^{n-1} e^{rq} \int_0^1 \overline{C}_m(s-t,q) f^{(m+1)}(x_r - sh + ht)\, dt \qquad (10.5)$$

$$= -\frac{h^{m+1}}{m!} \int_0^n \overline{C}_m(s-t,q) f^{(m+1)}(x_0 - sh + ht)\, dt. \qquad (10.6)$$

If we write $q + i\pi$ for q in (10.4), we obtain the formula for alternating sums of the form $\sum(-)^r e^{rq} f(x_r)$:

$$\sum_{r=0}^{n-1} (-)^r e^{rq} f(x_r) = \sum_{r=0}^{m} \frac{h^r C_r(s, q + i\pi)}{r!}$$

$$\times \{(-)^n e^{nq} f^{(r)}(x_n - sh) - f^{(r)}(x_0 - sh)\} + R(s, q + i\pi).$$

$$(10.7)$$

As noted at the beginning of Section 4 of Chapter I, the formulae for weighting functions which are polynomials in $\cos(\alpha_j x)$, $\sin(\beta_j x)$, $\cosh(a_j x)$, $\sinh(b_j x)$ and $e^{c_j x}$ $(j = 1, 2, 3, \ldots)$ may be obtained by forming appropriate linear combinations of (10.4) with $q = q_i$ $(i = 1, 2, 3, \ldots)$.

Thus we may obtain from (10.4) the summation formulae for the weighting functions $\cos(vx)$ and $\sin(vx)$ in the form

$$\sum_{r=0}^{n-1} w(\varepsilon + r) f(x_r) = \sum_{r=0}^{m} \frac{h^r}{r!} \{W_r(n) f^{(r)}(x_n - sh) - W_r(0) f^{(r)}(x_0 - sh)\} + R(s),$$

$$(10.8)$$

where ε is an arbitrary parameter and where the coefficient $W_r(k)$ $(k = 0, n)$ and the remainder $R(s)$ corresponding to the prescribed weighting functions are given in table (10.9), where $R(s, q)$ $(q = \pm iv)$ is given by (10.5 and 6).

Table (10.9)

$w(x)$	$W_r(k)$	$R(s)$
$\cos(vx)$	$\frac{1}{2}\{C_r(s, iv)\, e^{iv(k+\varepsilon)} + C_r(s, -iv)\, e^{-iv(k+\varepsilon)}\}$	$\frac{1}{2}\{e^{i\varepsilon v} R(s, iv) + e^{-i\varepsilon v} R(s, -iv)\}$
$\sin(vx)$	$-\dfrac{i}{2}\{C_r(s, iv)\, e^{iv(k+\varepsilon)} - C_r(s, -iv)\, e^{-iv(k+\varepsilon)}\}$	$-\dfrac{i}{2}\{e^{i\varepsilon v} R(s, iv) - e^{-i\varepsilon v} R(s, -iv)\}$

The formulae corresponding to the weighting functions $\cosh(ux)$ and $\sinh(ux)$, which may sometimes be useful in finite summation, may be obtained by writing iu for v in (10.8 and 9).

Since from (1.10 and 11) $e^{iv/2}C_r(\tfrac{1}{2}, iv)$ and $C_r(0, iv)$ are odd functions of v when r is even and even functions of v when r is odd (except for $C_0(0, iv)$ when $C_0(0, iv) + \tfrac{1}{2}$ is odd in v), the coefficients in the formula (10.8) assume simpler form when $s = 0$ or $s = \tfrac{1}{2}$.

Thus for $s = 0$ (10.8) assumes the form

$$\sum_{r=0}^{n} w(\varepsilon + r) f(x_r) = \tfrac{1}{2}\{w(\epsilon) f(x_0) + w(\varepsilon + n) f(x_n)\}$$

$$+ \sum_{r=0}^{m} \frac{h^r}{r!} \theta_r(iv) \{V_r(n + \varepsilon) f^{(r)}(x_n) - V_r(\varepsilon) f^{(r)}(x_0)\} + R(0),$$

$$(10.10)$$

where

$$\theta_0(iv) = C_0(0, iv) + \tfrac{1}{2}, \quad \theta_r(iv) = C_r(0, iv) \quad (r \geqslant 1) \tag{10.11}$$

and the form of the coefficient $V_r(y)$ corresponding to each weighting function is given by table (10.12).

Table (10.12)

$w(x)$	$V_r(y)$	
$\cos(vx)$	$i\sin(vy)$	$(r$ even$)$
	$\cos(vy)$	$(r$ odd$)$
$\sin(vx)$	$-i\cos(vy)$	$(r$ even$)$
	$\sin(vy)$	$(r$ odd$)$

Similarly for $s = \tfrac{1}{2}$ (10.8) reduces to

$$\sum_{r=0}^{n-1} w(\varepsilon + r) f(x_r) = \sum_{r=0}^{m} \frac{h^r}{r!} e^{iv/2} C_r(\tfrac{1}{2}, iv)$$

$$\times \{V_r(n + \varepsilon - \tfrac{1}{2}) f^{(r)}(x_n - h/2) - V_r(\varepsilon - \tfrac{1}{2}) f^{(r)}(x_0 - h/2)\} + R(\tfrac{1}{2}), \tag{10.13}$$

where, as before, the form of the coefficient $V_r(y)$ is given by (10.12).

11. *Error bounds for the formulae of Section* 10

It may be shown that (cf. Section 8)

$$\left| \int_0^1 \bar{C}_m(s - t, q) f^{(m+1)}(x_r - sh + ht)\, dt \right| \leqslant M\mu_r(q),$$

where

$$M = \max |e^{sq} C_m(s, q)| \quad (0 \leqslant s \leqslant 1) \tag{11.1}$$

and

$$\mu_r(q) = \max |e^{qt} f^{(m+1)}(x_r + ht)| \quad (-s \leqslant t \leqslant 1 - s). \tag{11.2}$$

Hence for $q = u + iv$ (10.5) yields

$$|R(s,q)| \leqslant \frac{h^{m+1}}{m!}\, M \sum_{r=0}^{n-1} \mathrm{e}^{ru}\, \mu_r(q) \tag{11.3}$$

$$\leqslant \begin{cases} \dfrac{h^{m+1}}{m!}\, M\mu^*(q)\, \dfrac{1-\mathrm{e}^{nu}}{1-\mathrm{e}^{u}} & (u \neq 0) \tag{11.4} \\[2ex] \dfrac{h^{m+1}}{m!}\, M\mu^*(q)\, n & (u = 0), \tag{11.5} \end{cases}$$

where

$$\mu^*(q) = \max\{\mu_r(q)\} \quad (0 \leqslant r \leqslant n-1). \tag{11.6}$$

Some upper bounds for M, we may note, are given in Section 5.

If $f^{(m+1)}(t)$ is of constant sign in $x_0 - sh \leqslant t \leqslant x_n - sh$ and if $q = iv$ is purely imaginary, then we have from (10.6)

$$|R(s,iv)| \leqslant \frac{h^m}{m!}\, M\, |f^{(m)}(x_n - sh) - f^{(m)}(x_0 - sh)|. \tag{11.7}$$

If $q = u$ is purely real, there follows from (5.21 and 22) that $C_r(s,u)$ retains its sign for $0 \leqslant s \leqslant 1$ and so also does $\bar{C}_r(s)$, for all s, by virtue of (10.1).

Therefore in the special case where $q = u$ is purely real we may apply the mean value theorem to (10.6) and obtain

$$R(s,u) = h^{m+1}(\mathrm{e}^{nu} - 1)C_{m+1}(s,u)f^{(m+1)}(\xi - sh)/(m+1)! \quad (x_0 < \xi < x_n). \tag{11.8}$$

If u is negative, there follows from (4.8) that in that case $C_r(s,u)$ is negative for all $s \geqslant 0$; so that if in addition it is given that $f^{(m+1)}(x)$ and $f^{(m+2)}(x)$ both retain their signs in $x_0 - sh \leqslant x \leqslant x_n - sh$ and are of opposite signs, then the error test applies and the error committed in truncating the series after retaining the mth term is in magnitude less than and is of the same sign as the $(m+1)$th term of the series.

If, on the other hand, u is positive, $C_r(s,u)$ has the sign $(-)^r$ for $s \leqslant 1$ and hence similar error test applies whenever $f^{(m+1)}(x)$ and $f^{(m+2)}(x)$ both retain their signs and are of the same sign in $x_0 - sh \leqslant x \leqslant x_n - sh$.

For the remainder terms in the formulae (10.8, 10 and 13) with the weighting functions $\cos(vx)$ and $\sin(vx)$ we may obtain from (11.3 and 5)

$$\tfrac{1}{2}|\mathrm{e}^{i\epsilon v} R(s,iv) \pm \mathrm{e}^{-i\epsilon v} R(s,-iv)| \leqslant \frac{h^{m+1}}{m!}\, M \sum_{r=0}^{n-1} \mu_r \tag{11.9}$$

$$\leqslant \frac{h^{m+1}}{m!}\, M\mu^* n, \tag{11.10}$$

where

$$\mu_r = \max |f^{(m+1)}(x_r + ht)| \ (-s \leqslant t \leqslant 1-s)$$

and

$$\mu^* = \max \{\mu_r\} \quad (0 \leqslant r \leqslant n-1).$$

If $f^{(m+1)}(t)$ retains its sign in $x_0 - sh \leqslant t \leqslant x_n - sh$, then we also have from (11.7)

$$\tfrac{1}{2} |e^{i\epsilon v} R(s, iv) \pm e^{-i\epsilon v} R(s, -iv)| \leqslant \frac{h^m}{m!} M |f^{(m)}(x_n - sh) - f^{(m)}(x_0 - sh)|. \quad (11.11)$$

It may be noticed that both the forms (11.5) and (11.10) will become indeterminate whenever the summation extends over an infinite number of terms.

CHAPTER III

OPERATIONAL METHOD

Introduction

In Chapters I and II we used complex variable and other methods to derive general rectangular formulae with prescribed weighting functions.

In the present chapter we consider the use of operators for the derivation of such general rectangular formulae with prescribed weighting functions and show that, starting from the formulation of the general problem in operational terms (Section 1), it is possible to devise operational procedures for the derivation of all the special formulae of Chapters I and II (Sections 3–5).

The utility of the operational approach is further exemplified by introducing in Section 2 the idea of a new type of formulae which feature arbitrarily chosen correction points—this idea is further developed later in Chapter IV.

1. *The general rectangular error operator*

For the purposes of the present chapter we shall require, in addition to the usual difference and differential operators, the fractional integral operator $D^{-\lambda}$ which may, consistently with the other operators, be defined as

$$D^{-\lambda}f(x) = \frac{1}{\Gamma(\lambda)} \int_a^x (x-t)^{\lambda-1} f(t)\,\mathrm{d}t, \tag{1.1}$$

where $\lambda > 0$ is not necessarily an integer, and a is a parameter suitably chosen so that each integral involved may exist.†

We also define I_n as the integral operator

$$I_n f(x) = \int_x^{x+nh} f(t)\,\mathrm{d}t. \tag{1.2}$$

---

† The choice of a as the lower limit is a slight departure from the usual definition of the integral operator via the convolution integral in operational calculus, where the lower limit is always chosen to be zero. The departure makes the convolution operation non-commutable. Since this is of no material disadvantage here, we prefer the parameter a, the proper choice of which ensures the existence of all the intergrals involved.

Then

$$DI_n = I_n D = E^n - 1. \tag{1.3}$$

Let

$$\Box_n f(x) \equiv h \sum_{r=0}^{n-1} f(x+sh+rh) - \int_x^{x+nh} f(t)\,\mathrm{d}t. \tag{1.4}$$

Then, since the right side of (1.4) denotes the error in approximating the integral of a function by its general n-ordinate rectangular sum, the operator $\Box_n$ may appropriately be called the *general rectangular error operator*.

On writing the right side of (1.4) in operational form and using (1.5), we may obtain the expressions

$$\Box_n = \left(\frac{hDE^s}{E-1} - 1\right) I_n \tag{1.5a}$$

$$= h\left(\frac{E^s}{E-1} - \frac{1}{hD}\right)(E^n - 1) \tag{1.5b}$$

$$= h\left(\frac{e^{shD}}{e^{hD}-1} - \frac{1}{hD}\right)(E^n - 1). \tag{1.5c}$$

We may now, with the help of these operational formulae (1.5a, b and c) above, expand $\Box_n$ differently to obtain the various correction formulae associated with the rectangular sum.

For example, since

$$\frac{e^{shD}}{e^{hD}-1} - \frac{1}{hD} = \sum_{r=0}^{\infty} \frac{B_{r+1}(s)}{(r+1)!}(hD)^r \tag{1.6}$$

we obtain, from (1.5c),

$$\Box_n f(x) = \sum_{r=0}^{\infty} h^{r+1} \frac{B_{r+1}(s)}{(r+1)!} D^r (E^n - 1) f(x)$$

$$= \sum_{r=0}^{\infty} h^{r+1} \frac{B_{r+1}(s)}{(r+1)!} \{f^{(r)}(x+nh) - f^{(r)}(x)\}. \tag{1.7}$$

On using this value of $\Box_n f(x)$ in (1.4) and writing x_0 for x, we obtain, finally,

$$h \sum_{r=0}^{n-1} f(x_r + sh) - \int_{x_0}^{x_n} f(t)\,\mathrm{d}t = \sum_{r=0}^{\infty} h^{r+1} \frac{B_{r+1}(s)}{(r+1)!} \{f^{(r)}(x_n) - f^{(r)}(x_0)\}, \tag{1.8}$$

which is the well-known general Euler–Maclaurin formula for the rectangular sum (Steffensen, 1950).

2. *The general rectangular formula with arbitrary correction points*

In the formula (1.8) we require the values of the function and its derivatives at the limits of the integral, i.e. it selects the limits of the integral as *correction points*. However, the availability of the derivatives may sometimes dictate the choice of the correction points. For such cases we require a correction formula where the correction points are open to choice, and this may be obtained operationally thus.

Equation (1.5c) may be written in the form

$$\square_n = h\left(\frac{e^{(s+\theta_n)hD}}{e^{hD}-1} - \frac{e^{\theta_n hD}}{hD}\right) E^{(n-\theta_n)}$$

$$- h\left(\frac{e^{(s+\theta_0)hD}}{e^{hD}-1} - \frac{e^{\theta_0 hD}}{hD}\right) E^{-\theta_0}, \tag{2.1}$$

where θ_n and θ_0 are any numbers; so that, expanding the terms in brackets in powers of hD (using (1.6)), $\square_n$ may be expressed as

$$\square_n = \left\{\sum_{r=0}^{\infty} h^{r+1}\frac{B_{r+1}(s+\theta_n)-(\theta_n)^{r+1}}{(r+1)!}D^r\right\} E^{(n-\theta_n)}$$

$$- \left\{\sum_{r=0}^{\infty} h^{r+1}\frac{B_{r+1}(s+\theta_0)-(\theta_0)^{r+1}}{(r+1)!}D^r\right\} E^{-\theta_0}. \tag{2.2}$$

Hence, operating on $f(x_0)$ and using (1.4), we obtain the required general formula in the form

$$h\sum_{r=0}^{n-1} f(x_r+sh) - \int_{x_0}^{x_n} f(t)\,\mathrm{d}t = \sum_{r=0}^{\infty} h^{r+1}\frac{B_{r+1}(s+\theta_n)-(\theta_n)^{r+1}}{(r+1)!}f^{(r)}(x_n-\theta_n h)$$

$$- \sum_{r=0}^{\infty} h^{r+1}\frac{B_{r+1}(s+\theta_0)-(\theta_0)^{r+1}}{(r+1)!}f^{(r)}(x_0-\theta_0 h), \tag{2.3}$$

where, θ_n and θ_0 being arbitrary, the correction points $(x_n-\theta_n h)$ and $(x_0-\theta_0 h)$ are now at our choice.

If we, for example, take $\theta_n = 1-s$ and $\theta_0 = -s$ we obtain a formula which has as correction points the two end ordinates used in the rectangular sum. On the other hand, a formula which is simpler to use may be obtained by taking $\theta_n = \theta_0 = -s$, giving correction points at (x_0+sh) and (x_n+sh), the last point being just outside the range of integration. The ensuing formulae are, perhaps, best presented in their pivotal form. This is obtained by a simple shift of origin, or, otherwise, operating from the start on $f(x_0 - sh)$, giving the two formulae

$$h \sum_{r=0}^{n-1} f(x_r) - \int_{x_0-sh}^{x_n-sh} f(t)\,\mathrm{d}t = \sum_{r=0}^{\infty} h^{r+1} \frac{B_{r+1}(1)-(1-s)^{r+1}}{(r+1)!} f^{(r)}(x_{n-1})$$

$$-\sum_{r=0}^{\infty} h^{r+1} \frac{B_{r+1}-(-s)^{r+1}}{(r+1)!} f^{(r)}(x_0) \qquad (2.4)$$

and

$$h \sum_{r=0}^{n-1} f(x_r) - \int_{x_0-sh}^{x_n-sh} f(t)\,\mathrm{d}t = \sum_{r=0}^{\infty} h^{r+1} \frac{B_{r+1}-(-s)^{r+1}}{(r+1)!} \{f^{(r)}(x_n)-f^{(r)}(x_0)\}, \quad (2.5)$$

where B_r is the rth Bernoulli number, and $B_r(1) = B_r\,(r>1)$, $B_1(1) = 1+B_1$.

Changing the sign of s in (2.5), we may also obtain the formula

$$h \sum_{r=0}^{n-1} f(x_r) - \int_{x_0+sh}^{x_n+sh} f(t)\,\mathrm{d}t = \sum_{r=0}^{\infty} h^{r+1} \frac{B_{r+1}-(s)^{r+1}}{(r+1)!} \{f^{(r)}(x_n)-f^{(r)}(x_0)\}. \quad (2.6)$$

The formulae (2.5) and (2.6) allow us, between them, to express the correction in terms of the values of the function and its derivatives at the pivotal points which are *the nearest* from the end points of the interval of integration—so that, in all cases, it is possible to have $0 \leqslant s \leqslant \tfrac{1}{2}$.

We may note here that this convenience of choosing the correction points is purchased at the price of having some of the coefficients larger.

It should also be noted that the general formula (2.3), of which the other formulae of this section are special cases, may also be obtained by the complex variable method, as well as by the method of integration by parts. These methods also provide us with expressions for the error committed in truncating the correction series after a finite number of terms. In Chapter IV these methods, as well as the operational method, are applied to obtain the arbitrary correction point formulae for the more general case in which there are prescribed weighting functions, which are unity in the formulae of the present section. An expression for the truncation error in the formula (2.3) is specially derived in Section 4 of Chapter IV.

In the remaining sections of this chapter attention is restricted to the operational methods for deriving the formulae, with correction points situated at the limits of the integral, which have been obtained in Chapters I and II by other methods.

3. *The general rectangular formula with the exponential weighting function*

Using the operational formula

$$\chi(D)\,\mathrm{e}^{px}f(x) = \mathrm{e}^{px}\chi(p+D)f(x), \tag{3.1}$$

where $\chi(D)$ is a suitable function of the differential operator D, we obtain from (1.5c)

$$\square_n\,\mathrm{e}^{px_0}f(x_0) = h(E^n-1)\,\mathrm{e}^{px_0}\left(\frac{\mathrm{e}^{sq}\,\mathrm{e}^{shD}}{\mathrm{e}^{q}\,\mathrm{e}^{hD}-1}-\frac{1}{q+hD}\right)f(x_0), \tag{3.2}$$

where, as before, $q = ph$.

But (cf. Chapter II, equation (3.2)),

$$\frac{\mathrm{e}^{sq}\,\mathrm{e}^{shD}}{\mathrm{e}^{q}\,\mathrm{e}^{hD}-1}-\frac{1}{q+hD} = \mathrm{e}^{sq}\sum_{r=0}^{\infty}\frac{D_r(s)}{r!}\,(hD)^r. \tag{3.3}$$

Hence

$$\square_n\,\mathrm{e}^{px_0}f(x_0) = \sum_{r=0}^{\infty}\frac{h^{r+1}\,\mathrm{e}^{sq}\,D_r(s)}{r!}\{\mathrm{e}^{px_n}f^{(r)}(x_n)-\mathrm{e}^{px_0}f^{(r)}(x_0)\}, \tag{3.4}$$

as derived before by the complex variable method and by the method of integration by parts (cf. Chapter I, Section 3 and Chapter II, Section 7).

4. *The weighting function* $(x-x_0)^{\alpha_0}$

Here the operation D^r on the weighting function $(x-x_0)^{\alpha_0}$ is, in general, not permissible at $x = x_0$, for any $r > \alpha_0$. It is therefore perhaps desirable not to think of the error operator $\square_n$ as a function of the differential operator D so long as the singular weighting function remains to the right of $\square_n$.

We may, instead, use the form

$$\square_n = h\left(\frac{E^s}{E-1}-\frac{1}{hD}\right)(E^n-1),$$

as given in (1.5b); and then

$$\square_n F(x_0) = h\left(\frac{E^s}{E-1}-\frac{1}{hD}\right)\{F(x_n)-F(x_0)\}, \tag{4.1}$$

where

$$F(x) \equiv (x-x_0)^{\alpha_0}f(x). \tag{4.2}$$

Consider† $\{E^s/(zE-1)\}\,F(x_0)$:

$$-\left(\frac{E^s}{zE-1}\right)F(x_0) = \left(\sum_{m=0}^{\infty} z^m\, E^{m+s}\right)F(x_0)$$

$$= \left\{\sum_{m=0}^{\infty} z^m\, h^{\alpha_0}(m+s)^{\alpha_0}\, E^{m+s}\right\}f(x_0) \quad \text{(from (4.2))}$$

$$= \left\{\sum_{m=0}^{\infty} h^{\alpha_0} z^m (m+s)^{\alpha_0} \sum_{r=0}^{\infty} \frac{(m+s)^r (hD)^r}{r!}\right\}f(x_0), \qquad (4.3)$$

since

$$E^{m+s} = \mathrm{e}^{(m+s)hD} = \sum_{r=0}^{\infty} \frac{(m+s)^r (hD)^r}{r!}.$$

Hence, inverting the order of the summations in (4.3), we have

$$-\left(\frac{E^s}{zE-1}\right)F(x_0) = \left\{h^{\alpha_0} \sum_{r=0}^{\infty} \frac{(hD)^r}{r!} \sum_{m=0}^{\infty} z^m (m+s)^{r+\alpha_0}\right\}f(x_0)$$

$$= \left\{h^{\alpha_0} \sum_{r=0}^{\infty} \frac{(hD)^r}{r!}\, \bar{\Phi}(z,\, -r-\alpha_0,\, s)\right\}f(x_0) \qquad (4.4)$$

$$\text{(cf. Appendix A, eq. (A1))}$$

$$= \left[h^{\alpha_0} \sum_{r=0}^{\infty} \frac{(hD)^r}{r!}\, z^{-s}\left\{\Gamma(r+1+\alpha_0)\left(\ln\frac{1}{z}\right)^{-(r+1+\alpha_0)}\right.\right.$$

$$\left.\left. +\, \zeta(-r-\alpha_0,\, s) + O(\ln z)\right\}\right]f(x_0), \qquad (4.5)$$

using the Lindelöf–Wirtinger expansion (cf. Appendix A, eq. (A5)).
But

$$\sum_{r=0}^{\infty} \frac{(hD)^r}{r!}\, \Gamma(r+1+\alpha_0)\left(\ln\frac{1}{z}\right)^{-(r+1+\alpha_0)} = \Gamma(1+\alpha_0)\left(\ln\frac{1}{z} - hD\right)^{-(1+\alpha_0)},$$

$$(4.6)$$

so that (4.5) reduces to

$$-\left(\frac{E^s}{zE-1}\right)F(x_0) = h^{\alpha_0} z^{-s}\left[\Gamma(1+\alpha_0)\left(\ln\frac{1}{z} - hD\right)^{-(1+\alpha_0)}\right.$$

$$\left. +\sum_{r=0}^{\infty}\left\{\frac{(hD)^r}{r!}\, \zeta(-r-\alpha_0,\, s) + O(\ln z)\right\}\right]f(x_0),$$

† For the analytical reasons for adopting the following procedure see Appendix B.

where we may now let z tend to 1 and obtain

$$\left(\frac{E^s}{E-1}\right) F(x_0) = \Big\{(-)^{-\alpha_0}\, \Gamma(1+\alpha_0)\, h^{-1}\, D^{-(1+\alpha_0)}$$

$$-h^{\alpha_0} \sum_{r=0}^{\infty} \frac{(hD)^r}{r!}\, \zeta(-r-\alpha_0, s)\Big\} f(x_0). \tag{4.7}$$

We also have, from (1.1),

$$\Gamma(1+\alpha_0)\,(-D)^{-(1+\alpha_0)} f(x_0) = -\int_a^{x_0} (t-x_0)^{\alpha_0} f(t)\, \mathrm{d}t$$

$$= (-D)^{-1} F(x_0), \tag{4.8}$$

the use of which in (4.7) above yields, finally, the expansion

$$\left(\frac{E^s}{E-1} - \frac{1}{hD}\right) F(x_0) = -\Big\{h^{\alpha_0} \sum_{r=0}^{\infty} \frac{(hD)^r}{r!}\, \zeta(-r-\alpha_0, s)\Big\} f(x_0). \tag{4.9}$$

Again, for operating on $F(x_n)$, the expansion

$$\frac{E^s}{E-1} - \frac{1}{hD} = \sum_{r=0}^{\infty} \frac{B_{r+1}(s)}{(r+1)!}\, (hD)^r, \tag{4.10}$$

as in (1.6), is appropriate.

Hence we obtain, using (4.9) and (4.10) in (4.1), the required correction formula

$$\Box_n F(x_0) = \sum_{r=0}^{\infty} \frac{h^{r+1} B_{r+1}(s)}{(r+1)!}\, F^{(r)}(x_n) + \sum_{r=0}^{\infty} \frac{h^{r+1+\alpha_0} \zeta(-r-\alpha_0, s)}{r!}\, f^{(r)}(x_0), \tag{4.11}$$

as derived before by the complex variable method (cf. Chapter I, Section 7).

5. *The weighting function* $e^{px}(x-x_0)^{\alpha_0}$

On performing the operation $(E^n - 1)$, we obtain from (3.2) writing E for e^{hD} and $F(x_0)$ for $f(x_0)$,

$$\Box_n e^{px_0} F(x_0) = h\, e^{px_n}\Big\{\frac{e^{sq} E^s}{e^q E - 1} - \frac{1}{h(p+D)}\Big\} F(x_n)$$

$$- h\, e^{px_0}\Big\{\frac{e^{sq} E^s}{e^q E - 1} - \frac{1}{h(p+D)}\Big\} F(x_0). \tag{5.1}$$

Again, since here, as in Section 4,

$$F(x) = (x-x_0)^{\alpha_0} f(x), \tag{5.2}$$

we obtain, taking $z = e^q$ in (4.4),

$$\left(\frac{E^s}{e^q E - 1}\right) F(x_0) = -h^{\alpha_0}\Big\{\sum_{r=0}^{\infty} \frac{(hD)^r}{r!}\, \Phi(e^q, -r-\alpha_0, s)\Big\} f(x_0). \tag{5.3}$$

Moreover, the lemma (4.8) may be generalized to

$$\Gamma(r+1)\,(-D)^{-(r+1)}\,F(x_0) = \Gamma(r+1+\alpha_0)\,(-D)^{-(r+1+\alpha_0)}f(x_0), \qquad (5.4)$$

so that

$$\left(\frac{1}{D+p}\right)F(x_0) = -\left\{\sum_{r=0}^{\infty} p^r(-D)^{-(r+1)}\right\}F(x_0)$$

$$= -\left\{\sum_{r=0}^{\infty} \frac{\Gamma(r+1+\alpha_0)}{\Gamma(r+1)}\,p^r(-D)^{-(r+1+\alpha_0)}\right\}f(x_0) \quad \text{(from (5.4),}$$

$$= \{(-)^{-\alpha_0}\,\Gamma(1+\alpha_0)\,(D+p)^{-(1+\alpha_0)}\}f(x_0)$$

$$= -\left\{\sum_{r=0}^{\infty} \frac{\Gamma(r+1+\alpha_0)}{\Gamma(r+1)}\,(-p)^{-(r+1+\alpha_0)}\,D^r\right\}f(x_0), \qquad (5.5)$$

expanding in ascending powers of D.

Hence, using (5.3) and (5.5), we obtain for the second group of terms in the right side of (5.1) the expansion

$$h\,\mathrm{e}^{px_0}\left\{\frac{\mathrm{e}^{sq}\,E^s}{\mathrm{e}^q\,E-1} - \frac{1}{h(p+D)}\right\}F(x_0)$$

$$= -\,\mathrm{e}^{px_0}\,\mathrm{e}^{sq}\left\{\sum_{r=0}^{\infty} h^{r+1+\alpha_0}\,\frac{G_r(s,q,\alpha_0)}{r!}\,D^r\right\}f(x_0), \qquad (5.6)$$

where

$$G_r(s,q,\alpha_0) = \Phi(\mathrm{e}^q,\,-r-\alpha_0,\,s) - \Gamma(r+1+\alpha_0)\,(-q)^{-(r+1+\alpha_0)}\,\mathrm{e}^{-sq}. \qquad (5.7)$$

Again, for the first group of terms in the right side of (5.1) the expansion

$$\frac{\mathrm{e}^{sq}\,\mathrm{e}^{shD}}{\mathrm{e}^q\,\mathrm{e}^{hD}-1} - \frac{1}{q+hD} = \mathrm{e}^{sq}\sum_{r=0}^{\infty} \frac{D_r(s)}{r!}\,(hD)^r, \qquad (5.8)$$

as in (3.3), is appropriate.

Hence, using (5.8) and (5.6) in (5.1), and from (1.4) we finally obtain the required formula:

$$h\sum_{r=0}^{n-1}(rh+sh)^{\alpha_0}\,\mathrm{e}^{p(x_r+sh)}f(x_r+sh) - \int_{x_0}^{x_n}(x-x_0)^{\alpha_0}\,\mathrm{e}^{px}f(x)\,\mathrm{d}x$$

$$= \sum_{r=0}^{\infty} \frac{h^{r+1}\,\mathrm{e}^{sq}\,D_r(s)}{r!}\,\mathrm{e}^{px_n}\,F^{(r)}(x_n)$$

$$+ \sum_{r=0}^{\infty} \frac{h^{r+1+\alpha_0}\,\mathrm{e}^{sq}\,G_r(s,q,\alpha_0)}{r!}\,\mathrm{e}^{px_0}f^{(r)}(x_0), \qquad (5.9)$$

as obtained also in Appendix A, by using the complex variable method.

For the properties and the formulae for the computation of the **function** $\Phi(\mathrm{e}^q,\,-r-\alpha_0,\,s)$ featuring in (5.7), reference may be made to *Higher transcendental functions*, vol. I.

CHAPTER IV

ARBITRARY CORRECTION POINT FORMULAE AND FORMULAE WITH CORRECTION TERMS EXPRESSED IN TERMS OF QUANTITIES OTHER THAN DERIVATIVES

Introduction

In Chapter III (Section 2) we obtained the general rectangular formula with arbitrarily prescribed correction points for the unit weighting function by operational method. A brief but, it is hoped, adequate treatment is given below (Sections 1–4) of the general problem of deriving such arbitrary correction point formulae with prescribed weighting functions and also of the allied problem of truncation error.

In the remaining sections of the present chapter attention is given to the formulae which express the correction terms in terms of quantities other than derivatives. The explicit form of the coefficients are derived for the forward, backward and central difference formulae for (1) the unit weighting function, (2) the exponential and allied weighting functions and (3) the weighting function

$$w(x) = (x - x_0)^{\alpha_0} (x_n - x)^{\alpha_n}.$$

The use of these formulae is exemplified with the help of several numerical examples, and short tables of the coefficients for the trapezoidal case of (3) are given in Appendix C.

Also given are the rational fractional values of the first eight coefficients of the mid-ordinate formula (6.7) which, it may be noticed, fills an obvious gap in the usual list of formulae for numerical integration. I wish this formula to be known as *Bickley's formula*.

1. *The derivation of arbitrary correction point formulae:* *(i) Complex variable method*

In Section 2 of Chapter I we obtained the general rectangular formula with the correction points situated at the limits of the integral, $x_k \, (k = 0, n)$, by expanding $f_k(x_k + iht/2\pi)$ in a Taylor series centred at x_k. We may, instead, expand $f_k(x_k + iht/2\pi)$ in a Taylor series centred at $x_k - \theta_k h$ and obtain, in place of (2.5) of Chapter I, the expansion

$$F(x_k + iht/2\pi) = w_k\left(x_k + \frac{iht}{2\pi}\right)\left\{\sum_{r=0}^{m_k} \frac{h^r(\theta_k + it/2\pi)^r}{r!} f_k^{(r)}(x_k - \theta_k h)\right.$$

$$\left. + R_{m_k}^*(x_k + iht/2\pi)\right\}. \quad (1.1)$$

If we now replace (2.5) of Chapter I by (1.1) above and continue with the rest of the analysis in Section 2 of Chapter I, we obtain, writing

$$B_{rk}(\xi, s) = \sum_{j=0}^{r} \binom{r}{j}(-2\pi i\theta_k)^{r-j} A_{jk}(\xi, s), \quad (1.2)$$

where $A_{jk}(\xi, s)$ is given by (2.9) of Chapter I, the general rectangular formula for the same weighting functions as prescribed in Section 2 of Chapter I, but with the arbitrary correction points $x_n - \theta_n h$ and $x_0 - \theta_0 h$, in the form

$$h\sum_{r=0}^{n-1} F(x_r + sh) - \int_{x_0}^{x_n} F(x)\,dx$$

$$= \sum_{r=0}^{m_n} \frac{\psi_r^n(h, s)}{r!} f_n^{(r)}(x_n - \theta_n h) - \sum_{r=0}^{m_0} \frac{\psi_r^0(h, s)}{r!} f_0^{(r)}(x_0 - \theta_0 h) + R^*, \quad (1.3)$$

where

$$\psi_r^k(h, s) = \left(\frac{ih}{2\pi}\right)^{r+1} B_{rk}(0, s), \quad (1.4)$$

$$R^* = \sigma_n^*(s) + \rho_n^*(s) - \{\sigma_0^*(s) + \rho_0^*(s)\} + J(s), \quad (1.5)$$

$$\rho_k^*(s) = \frac{ih}{2\pi}\int_0^{2\pi b/h}\left\{\frac{w_k(x_k + iht/2\pi) R_{m_k}^*(x_k + iht/2\pi)}{1 - e^{2\pi is}\,e^t}\right.$$

$$\left. - \frac{w_k(x_k - iht/2\pi) R_{m_k}^*(x_k - iht/2\pi)}{1 - e^{-2\pi is}\,e^t}\right\}dt, \quad (1.6)$$

$$\sigma_k^*(s) = -\sum_{r=0}^{m_k}\left(\frac{ih}{2\pi}\right)^{r+1}\frac{B_{rk}(2\pi b/h, s)}{r!} f_k^{(r)}(x_k - \theta_k h) \quad (1.7)$$

and $J(s)$ as given before by (2.3) of Chapter I.

Using (2.12) of Chapter I and (1.2), (1.4) may be written in the form

$$\psi_r^k(h, s) = \sum_{j=0}^{r}\binom{r}{j}(\theta_k h)^{r-j}\phi_j^k(h, s)$$

$$= \{\phi^k(h, s) + \theta_k h\}^r \quad (1.8)$$

symbolically, where, after expansion, each $\{\phi^k(h, s)\}^j$ is to be replaced by $\phi_j^k(h, s)$.

5

Thus the coefficients $\psi_r^k(h, s)$ for an arbitrary correction point formula with prescribed weighting functions are simple linear combinations of the coefficients $\phi_j^k(h, s)$ of the corresponding formula (2.11) of Chapter I with the same weighting functions, but where the correction points are situated at the limits of the integral.

2. (*ii*) *Operational method*

The rule for calculating the correction coefficients may also be obtained operationally.

Suppose for the prescribed weighting functions, i.e. for

$$F(x) = w_k(x)f_k(x) \text{ at } x = x_k \quad (k = 0, \text{ or } n), \tag{2.1}$$

we have (see (1.4) of Chapter III)

$$\square_n F(x_0) = \left\{ \sum_{r=0}^{\infty} \frac{\phi_r^n(h, s)}{r!} D^r \right\} f_n(x_n) - \left\{ \sum_{r=0}^{\infty} \frac{\phi_r^0(h, s)}{r!} D^r \right\} f_0(x_0) \tag{2.2}$$

for the formula where the correction points are situated at the limits of the integral.

Then, writing

$$f_k(x_k) = e^{\theta_k h D} f_k(x_k - \theta_k h) \quad (k = 0, n) \tag{2.3}$$

in (2.2), we immediately obtain the required expansion for a formula with the same weighting functions but with arbitrary correction points in the form

$$\square_n F(x_0) = \left\{ \sum_{r=0}^{\infty} \frac{\psi_r^n(h, s)}{r!} D^r \right\} f_n(x_n - \theta_n h) - \left\{ \sum_{r=0}^{\infty} \frac{\psi_r^0(h, s)}{r!} D^r \right\} f_0(x_0 - \theta_0 h), \tag{2.4}$$

where there follows from (2.2 and 3) that $\psi_r^k(h, s)$ is given by (1.8).

3. (*iii*) *The method of integration by parts*

Let, in the notation of (1.8) above,

$$R_k = - \frac{1}{(m_k)!} \int_{-\theta_k h}^{0} \{\phi^k(h, s) - t\}^{m_k} f_k^{(m_k+1)}(x_k + t) \, dt. \tag{3.1}$$

Then, by successive integration by parts,

$$R_k = \sum_{r=0}^{m_k} \frac{\{\phi^k(h, s) + \theta_k h\}^r}{r!} f_k^{(r)}(x_k - \theta_k h) - \sum_{r=0}^{m_k} \frac{\phi_r^k(h, s)}{r!} f_k^{(r)}(x_k). \tag{3.2}$$

Hence, if we have for the prescribed weighting functions the formula

$$h \sum_{r=0}^{n-1} F(x_r + sh) - \int_{x_0}^{x_n} F(x)\,\mathrm{d}x = \sum_{r=0}^{m_n} \frac{\phi_r^n(h,s)}{r!} f_n^{(r)}(x_n)$$

$$- \sum_{r=0}^{m_0} \frac{\phi_r^0(h,s)}{r!} f_0^{(r)}(x_0) + R, \qquad (3.3)$$

where the correction points are situated at the limits of the integral, and R is the remainder due to the truncation of the correction series, then, using (3.2) to shift the correction points, we obtain, from (3.3), the formula with the same weighting functions, but with arbitrary correction points in the form

$$h \sum_{r=0}^{n-1} F(x_r + sh) - \int_{x_0}^{x_n} F(x)\,\mathrm{d}x = \sum_{r=0}^{m_n} \frac{\{\phi^n(h,s) + \theta_n\, h\}^r}{r!} f_n^{(r)}(x_n - \theta_n h)$$

$$- \sum_{r=0}^{m_0} \frac{\{\phi^0(h,s) + \theta_0\, h\}^r}{r!} f_0^{(r)}(x_0 - \theta_0 h) + R^*,$$

$$(3.4)$$

where

$$R^* = R + R_0 - R_n. \qquad (3.5)$$

From (3.5) it is seen that the truncation error R^* is composed of the remainder R of a similarly truncated formula with the same weighting functions, but with end point correction, and the additional error $R_0 - R_n$ which arises due to the shift in these correction points.

4. The general rectangular formula with arbitrary correction points for the unit weighting function

Since the general formulae obtained in the preceding sections (Sections 1–3) are all sufficiently explicit, it is not deemed necessary to provide individual treatment for each special weighting function considered in the earlier chapters. As an illustration, however, we apply the formulae of Section 3 to rederive the formula (2.3) of Chapter III, this time with an expression for the relevant truncation error.

In this case, for $k = 0$ and n, $w_k(x) = 1$, $f_k(x) = f(x)$, and

$$\phi_r^k(h,s) = \frac{h^{r+1} B_{r+1}(s)}{r+1} \qquad (4.1)$$

so that

$$\{\phi^k(h,s) + \theta_k h\}^r = \sum_{j=0}^{r} \binom{r}{j} \frac{h^{r+1} B_{j+1}(s)\, \theta_k^{r-j}}{j+1}$$

$$= h^{r+1} \frac{B_{r+1}(s + \theta_k) - \theta_k^{r+1}}{r+1}. \qquad (4.2)$$

5*

Hence, for $m_0 = m_n = m$, we obtain from (3.4) the required formula in the form

$$h \sum_{r=0}^{n-1} f(x_r + sh) - \int_{x_0}^{x_n} f(x)\, \mathrm{d}x$$

$$= \sum_{r=0}^{m} \frac{h^{r+1}}{(r+1)!} \{B_{r+1}(s+\theta_n) - (\theta_n)^{r+1}\} f^{(r)}(x_n - \theta_n h)$$

$$- \sum_{r=0}^{m} \frac{h^{r+1}}{(r+1)!} \{B_{r+1}(s+\theta_0) - (\theta_0)^{r+1}\} f^{(r)}(x_0 - \theta_0 h) + R^*, \quad (4.3)$$

where $R^* = R + R_0 - R_n$ is composed of

(i) R, the remainder in a similarly truncated general Euler–Maclaurin formula (see (1.8) of Chapter III), which is usually expressed as (Steffensen, 1950; Nörlund, 1924)

$$R = - \frac{h^{m+2}}{(m+1)!} \int_0^n \bar{B}_{m+1}(s-t) f^{(m+1)}(x_0 + ht)\, \mathrm{d}t, \qquad (4.4)$$

where $\bar{B}_r(z)$ is the periodic Bernoulli function, and

(ii) the 'shift error' $R_0 - R_n$, where, from (4.2) and (3.1),

$$R_k = - \frac{h^{m+2}}{(m+1)!} \int_{-\theta_k}^{0} \{B_{m+1}(s-t) - (-t)^{m+1}\} f^{(m+1)}(x_k + ht)\, \mathrm{d}t. \quad (4.5)$$

5. *The derivation of formulae with correction terms expressed in terms of quantities other than derivatives*

We have so far considered formulae where the correction terms are expressed in terms of the values of the function

$$f_k(x) = F(x)/w_k(x) \quad (k = 0, n)$$

and its derivatives at the corresponding correction point. These derivatives may be obtainable analytically if $f_k(x)$ is given by a sufficiently simple analytical formula or if it satisfies a simple enough differential equation; and in other cases these have to be computed with the help of some suitably chosen numerical formulae, a large selection of which is available in the literature. In all these cases use may also be made of the arbitrary correction point formulae of Sections 1 to 4 to shift the correction points to more convenient positions adjacent to the end points of the interval of integration.

A different approach is to transform the correction formula itself, prior to its application, by replacing the derivatives by the formal substitution of their equivalents in terms of other quantities, such as differences, or tabular values, ... (of the function, or of its first derivative, ...).

For example, the use of the formula-pair

$$(hD)^r = r! \sum_{\nu=r}^{\infty} S_\nu^r \frac{\Delta^\nu}{\nu!} = (-)^r r! \sum_{\nu=r}^{\infty} (-)^\nu S_\nu^r \frac{\nabla^\nu}{\nu!}, \tag{5.1}$$

where S_ν^r denote Stirling numbers of the first kind, enables us to express the correction terms in any formula of the preceding Chapters I, II and III (these are all of the form (3.3)) in terms of the forward or the backward differences of $f_k(x)$ at x_k ($k = 0, n$).

In forming these differences, however, we are unable, except in the special case when $s = 0$, to utilize the available values of the integrand at the pivotal points $x = x_r + sh$ ($r = 0, 1, \ldots, n-1$) which are used in forming the associated rectangular sum; and consequently we may wish to derive formulae which involve the differences of $f_0(x)$ and $f_n(x)$ at suitably chosen pivotal points in the neighbourhoods of x_0 and x_n respectively.

These formulae may be obtained by replacing the derivatives in the formula by using, instead of (5.1), the formulae

$$D^r f_k(x_k) \begin{cases} = \dfrac{1}{h^r}\left[S_r^r + \left(S_r^r \theta_k + \dfrac{S_{r+1}^r}{r+1} \right) \Delta + \left\{ S_r^r \dfrac{\theta_k(\theta_k-1)}{2!} \right. \\ \qquad \left. + \dfrac{S_{r+1}^r}{r+1}\theta_k + \dfrac{S_{r+2}^r}{(r+1)(r+2)} \right\} \Delta^2 + \ldots \right] \Delta^r f_k(x_k - \theta_k h) \\[2ex] = \dfrac{1}{h^r}\left[S_r^r - \left(-S_r^r \theta_k + \dfrac{S_{r+1}^r}{r+1} \right) \nabla + \left\{ S_r^r \dfrac{\theta_k(\theta_k+1)}{2!} \right. \\ \qquad \left. - \dfrac{S_{r+1}^r}{r+1}\theta_k + \dfrac{S_{r+2}^r}{(r+1)(r+2)} \right\} \nabla^2 - \ldots \right] \nabla^r f_k(x_k - \theta_k h), \end{cases} \tag{5.2}$$

where, for $k = 0$ and $k = n$, θ_k is so chosen that $x_0 - \theta_0 h$ and $x_n - \theta_n h$ coincide with the chosen pivotal points.

Alternatively, we may first shift the correction points to the chosen pivotal points by appropriately choosing θ_0 and θ_n in the arbitrary correction point formula of form (3.4) corresponding to the given formula, followed by the use of (5.1) to express the derivatives in terms of forward or backward differences.

While both the procedures outlined above are perfectly general, the latter, we may note, is likely to be more frequently useful since formulae of the form (5.1) for numerical differentiation at a pivotal point are usually far simpler in form compared with the corresponding formulae of the form (5.2) for the derivatives at an arbitrary point and also occur more often in the literature.

6. *Operational method: The general rectangular formula with forward and/or backward difference correction*

In some cases the operational method may be used.

For example, in the case when

$$w_n(x) = w_0(x) = 1 \quad \text{and} \quad f_n(x) = f_0(x) = f(x),$$

the formulae with the correction terms expressed in terms of the forward and/or backward differences (instead of derivatives as in Section 2 of Chapter III) of $f(x)$ at the arbitrary correction points $x_n - \theta_n h$ and $x_0 - \theta_0 h$ may be obtained by writing formula (1.5b) of Chapter III in the form

$$\Box_n = h\Omega(\theta_n)\, E^{n-\theta_n} - h\Omega(\theta_0)\, E^{-\theta_0},$$

where, for $k = 0$ and n, $\Omega(\theta_k)$ has the expressions

$$\Omega(\theta_k) = \frac{(1+\Delta)^{s+\theta_k}}{\Delta} - \frac{(1+\Delta)^{\theta_k}}{\ln(1+\Delta)} = \frac{(1-\nabla)^{1-s-\theta_k}}{\nabla} + \frac{(1-\nabla)^{-\theta_k}}{\ln(1-\nabla)}$$

(since $E = 1+\Delta = (1-\nabla)^{-1}$), and subsequently expanding $\Omega(\theta_k)$ in powers of Δ or ∇ as required.

Writing

$$\frac{(1+t)^y}{\ln(1+t)} = 1/t + \sum_{r=0}^{\infty} \beta_{r+1}(y)\, t^r, \tag{6.1}$$

so that

$$\beta_r(y) = \binom{y}{r} b_0 + \binom{y}{r-1} b_1 + \ldots + b_r, \tag{6.2}$$

where the coefficients $b_0 = 1$, $b_1 = \frac{1}{2}$, $b_2 = -\frac{1}{12}, \ldots$ are given (the first ten coefficients are listed in Jordan, 1939, pp. 265 et seq., where the properties of $\beta_r(y)$ are also discussed) by the relation

$$t/\ln(1+t) = \sum_{r=0}^{\infty} b_r\, t^r, \tag{6.3}$$

these expansions may be obtained in the forms

$$\Omega(\theta_k) = \sum_{r=0}^{\infty} \left\{ \binom{s+\theta_k}{r+1} - \beta_{r+1}(\theta_k) \right\} \Delta^r \tag{6.4}$$

$$= \sum_{r=0}^{\infty} (-)^r \left\{ \beta_{r+1}(-\theta_k) - \binom{1-s-\theta_k}{r+1} \right\} \nabla^r. \tag{6.5}$$

Hence, using (1.4) of Chapter III, the required group of formulae may be written in the form

$$h \sum_{r=0}^{n-1} f(x_r + sh) - \int_{x_0}^{x_n} f(x)\,\mathrm{d}x$$

$$= \begin{cases} h \sum_{r=0}^{\infty} \left\{ \binom{s+\theta_n}{r+1} - \beta_{r+1}(\theta_n) \right\} \Delta^r f(x_n - \theta_n h) \\ \text{or} \\ h \sum_{r=0}^{\infty} (-)^r \left\{ \beta_{r+1}(-\theta_n) - \binom{1-s-\theta_n}{r+1} \right\} \nabla^r f(x_n - \theta_n h) \end{cases}$$

$$- \begin{cases} h \sum_{r=0}^{\infty} \left\{ \binom{s+\theta_0}{r+1} - \beta_{r+1}(\theta_0) \right\} \Delta^r f(x_0 - \theta_0 h) \\ \text{or} \\ h \sum_{r=0}^{\infty} (-)^r \left\{ \beta_{r+1}(-\theta_0) - \binom{1-s-\theta_0}{r+1} \right\} \nabla^r f(x_0 - \theta_0 h) \end{cases}. \qquad (6.6)$$

The special cases $\theta_k = -s$ and $\theta_k = 1-s$ ($k = 0$ or n), which correspond respectively to correction points at the nearest pivotal points to the right and to the left of x_k, are of importance.

Thus, if we select the ∇-expansion at $x_n - \theta_n h$ and the Δ-expansion at $x_0 - \theta_0 h$, (6.6) yields, for $-\theta_0 = s = \frac{1}{2} = \theta_n$, the formula

$$h \sum_{r=0}^{n-1} f(x_r + h/2) - \int_{x_0}^{x_n} f(x)\,\mathrm{d}x$$

$$= h \sum_{r=0}^{\infty} (-)^r \beta_{r+1}(-\tfrac{1}{2}) \left\{ \nabla^r f(x_n - h/2) + (-)^r \Delta^r f(x_0 + h/2) \right\}$$

$$= -\frac{h}{24}(\nabla f_{n-\frac{1}{2}} - \Delta f_{\frac{1}{2}}) - \frac{h}{24}(\nabla^2 f_{n-\frac{1}{2}} + \Delta^2 f_{\frac{1}{2}})$$

$$-\frac{223h}{5760}(\nabla^3 f_{n-\frac{1}{2}} - \Delta^3 f_{\frac{1}{2}}) - \frac{103h}{2880}(\nabla^4 f_{n-\frac{1}{2}} + \Delta^4 f_{\frac{1}{2}})$$

$$-\frac{32119h}{967680}(\nabla^5 f_{n-\frac{1}{2}} - \Delta^5 f_{\frac{1}{2}}) - \frac{1111h}{35840}(\nabla^6 f_{n-\frac{1}{2}} + \Delta^6 f_{\frac{1}{2}})$$

$$-\frac{13528301h}{464486400}(\nabla^7 f_{n-\frac{1}{2}} - \Delta^7 f_{\frac{1}{2}})$$

$$-\frac{3194621h}{116121600}(\nabla^8 f_{n-\frac{1}{2}} + \Delta^8 f_{\frac{1}{2}}) - \dots. \qquad (6.7)$$

This formula, which I wish to be known as *Bickley's formula*, expresses the correction terms associated with the *mid-ordinate rule* in terms of the forward and the backward differences of $f(x)$ evaluated, respectively, at the first and the last pivotal points used in forming the mid-ordinate sum.

If no differences beyond the $(n-1)$th are retained, only the values of the integrand used in forming the mid-ordinate sum are involved.

7. *The forward and backward difference formulae for the exponential and allied weighting functions*

The relationship (3.2) of Chapter III may be written in the form

$$\square_n \, e^{px_0} f(x_0) = h \, e^{px_n} \Omega(\theta_n) f(x_n - \theta_n h) - h \, e^{px_0} \Omega(\theta_0) f(x_0 - \theta_0 h), \quad (7.1)$$

where, for $k = 0$ and n, $\Omega(\theta_k)$ has the expressions

$$\Omega(\theta_k) = \frac{e^{sq}(1+\Delta)^{s+\theta_k}}{e^q(1+\Delta)-1} - \frac{(1+\Delta)^{\theta_k}}{q+\ln(1+\Delta)} \quad (7.2)$$

$$= \frac{e^{sq}(1-\nabla)^{1-s-\theta_k}}{e^q-(1-\nabla)} - \frac{(1-\nabla)^{-\theta_k}}{q-\ln(1-\nabla)}, \quad (7.3)$$

which, for $\theta_n = \theta_0 = s = 0$, yield the expansions

$$\Omega(\theta_k)\Big|_{\substack{s=0 \\ \theta_k=0}} = \frac{1}{e^q-1} - \frac{1}{q} + \sum_{r=1}^{\infty} \left\{ \frac{(-)^r e^{-q}}{(1-e^{-q})^{r+1}} + g_r(q) \right\} \Delta^r \quad (7.4)$$

$$= \frac{1}{e^q-1} - \frac{1}{q} + \sum_{r=1}^{\infty} (-)^r \left\{ \frac{e^q}{(e^q-1)^{r+1}} - g_r(-q) \right\} \nabla^r, \quad (7.5)$$

where

$$g_r(q) = \frac{1}{r!} \sum_{j=1}^{r} j! \, S_r^j (-1/q)^{j+1} \quad (r \geqslant 1). \quad (7.6)$$

Hence, using (1.4) of Chapter III, the *trapezoidal formula* with the exponential weighting function and with the correction terms expressed in terms of the forward and/or the backward differences may be written in the form

$$h\left\{ \tfrac{1}{2} e^{px_0} f(x_0) + \sum_{r=1}^{n-1} e^{px_r} f(x_r) + \tfrac{1}{2} e^{px_n} f(x_n) \right\} - \int_{x_0}^{x_n} e^{px} f(x) \, dx$$

$$= h\chi_0(q) \left\{ e^{px_n} f(x_n) - e^{px_0} f(x_0) \right\}$$

$$+ h \, e^{px_n} \left\{ \begin{matrix} \displaystyle\sum_{r=1}^{\infty} \chi_r(q) \, \Delta^r f(x_n) \\[1mm] \text{or} \\[1mm] \displaystyle\sum_{r=1}^{\infty} (-)^{r+1} \chi_r(-q) \, \nabla^r f(x_n) \end{matrix} \right\}$$

$$- h \, e^{px_0} \left\{ \begin{matrix} \displaystyle\sum_{r=1}^{\infty} \chi_r(q) \, \Delta^r f(x_0) \\[1mm] \text{or} \\[1mm] \displaystyle\sum_{r=1}^{\infty} (-)^{r+1} \chi_r(-q) \, \nabla^r f(x_0) \end{matrix} \right\}, \quad (7.7)$$

where

$$\chi_0(q) = \frac{1}{2}\left(\frac{e^q+1}{e^q-1}\right) - \frac{1}{q}, \quad \chi_r(q) = \frac{(-)^r e^{-q}}{(1-e^{-q})^{r+1}} + g_r(q) \quad (r \geqslant 1). \tag{7.8}$$

If, on the other hand, we derive (7.7) by replacing the derivatives in (3.14) of Chapter I with the help of (5.1), we obtain the alternative formulae

$$\chi_0(q) = D_0(0,q) + \tfrac{1}{2}, \quad \chi_r(q) = \frac{1}{r!}\sum_{\nu=1}^{r} S_r^\nu D_\nu(0,q) \quad (r \geqslant 1), \tag{7.9}$$

which express $\chi_r(q)$ in terms of the Stirling numbers of the first kind and the D-functions of Chapter II, and are particularly useful when $|q|$ is small, since in that case both the power series expansion (4.10) of Chapter II and the asymptotic formula in Section 6 of Chapter II are available for the computation of $D_r(0,q)$.

By the direct substitution of (4.10) of Chapter II in (7.9), we may also obtain the expansion of $\chi_r(q)$ as a series in ascending powers of q in the form

$$\chi_0(q) = \sum_{j=0}^{\infty} \frac{B_{j+2}}{(j+2)!} q^{j+1}, \quad \chi_r(q) = \frac{1}{r!}\sum_{j=0}^{\infty} c_j^r q^j \quad (r \geqslant 1), \tag{7.10}$$

where

$$c_j^r = \sum_{\nu=1}^{r} \frac{S_r^\nu B_{j+\nu+1}}{(j+\nu+1)j!}. \tag{7.11}$$

The formulae for the weighting functions $\cosh(\alpha x)$, $\sinh(\alpha x)$, $\cos(\beta x)$ and $\sin(\beta x)$ may be obtained by forming appropriate linear combinations of (7.7) with $p = \pm\alpha, \pm i\beta$, and are of the form

$$\frac{h}{2} w(x_0)f(x_0) + h\sum_{r=1}^{n-1} w(x_r)f(x_r) + \frac{h}{2} w(x_n)f(x_n) - \int_{x_0}^{x_n} w(x)f(x)\,dx$$

$$= \left\{\begin{array}{c} h\sum_{r=0}^{\infty} U_r(x_n)\Delta^r f(x_n) \\ \text{or} \\ h\sum_{r=0}^{\infty} V_r(x_n)\nabla^r f(x_n) \end{array}\right\} - \left\{\begin{array}{c} h\sum_{r=0}^{\infty} U_r(x_0)\Delta^r f(x_0) \\ \text{or} \\ h\sum_{r=0}^{\infty} V_r(x_0)\nabla^r f(x_0) \end{array}\right\}, \tag{7.12}$$

where, using the notations

$$\alpha h = u, \quad \beta h = v, \quad \chi_r^{\mathrm{I}}(q) = \tfrac{1}{2}\{\chi_r(q) + \chi_r(-q)\},$$

$$\chi_r^{\mathrm{II}}(q) = \tfrac{1}{2}\{\chi_r(q) - \chi_r(-q)\}, \quad g_r^{\mathrm{I}}(q) = \tfrac{1}{2}\{g_r(q) + g_r(-q)\} \tag{7.13}$$

and

$$g_r^{\mathrm{II}}(q) = \tfrac{1}{2}\{g_r(q) - g_r(-q)\},$$

the coefficients $U_r(x_k)$ and $V_r(x_k)$ $(k = 0, n)$, corresponding to each weighting function, are given in tables (7.14), (7.15) and (7.16).

$$\text{Table (7.14)}$$

$w(x)$	$U_0(x_k) = V_0(x_k)$
$\cosh(\alpha x)$	$\sinh(\alpha x_k)\,\chi_0(u) = \tfrac{1}{2}\sinh(\alpha x_k)\{\coth(u/2) - 2/u\}$
$\sinh(\alpha x)$	$\cosh(\alpha x_k)\,\chi_0(u) = \tfrac{1}{2}\cosh(\alpha x_k)\{\coth(u/2) - 2/u\}$
$\cos(\beta x)$	$i\sin(\beta x_k)\,\chi_0(iv) = \tfrac{1}{2}\sin(\beta x_k)\{\cot(v/2) - 2/v\}$
$\sin(\beta x)$	$-i\cos(\beta x_k)\,\chi_0(iv) = -\tfrac{1}{2}\cos(\beta x_k)\{\cot(v/2) - 2/v\}$

It may be noted here that the formulae for the more general case of

$$\int_{x_0}^{x_n} w(x+\varepsilon)f(x)\,\mathrm{d}x,$$

which may be obtained from (7.12) by a change of variable, are of the form

$$h\left\{\tfrac{1}{2}w(x_0+\varepsilon)f(x_0) + \sum_{r=1}^{n-1} w(x_r+\varepsilon)f(x_r) + \tfrac{1}{2}w(x_n+\varepsilon)f(x_n)\right\}$$

$$-\int_{x_0}^{x_n} w(x+\varepsilon)f(x)\,\mathrm{d}x$$

$$= \begin{cases} h\sum_{r=0}^{\infty} U_r(x_n+\varepsilon)\,\Delta^r f(x_n) \\ \text{or} \\ h\sum_{r=0}^{\infty} V_r(x_n+\varepsilon)\,\nabla^r f(x_n) \end{cases} - \begin{cases} h\sum_{r=0}^{\infty} U_r(x_0+\varepsilon)\,\Delta^r f(x_0) \\ \text{or} \\ h\sum_{r=0}^{\infty} V_r(x_0+\varepsilon)\,\nabla^r f(x_0) \end{cases}, \quad (7.17)$$

where the coefficients $U_r(x_k+\varepsilon)$ and $V_r(x_k+\varepsilon)$ $(k=0,n)$, corresponding to each of the weighting functions of this section, are to be obtained by writing $x_k+\varepsilon$ for x_k in the tables (7.14 to 16) for $U(x_k)$ and $V(x_k)$.

8. *The central difference formulae for the exponential and allied weighting functions*

The central difference formulae for derivatives may be written in the forms

$$D^r \begin{cases} = \dfrac{1}{h^r} \sum_{\nu=0}^{\infty} a_{2\nu+r}^r\, \delta^{2\nu+r} & (8.1) \\[2ex] = \dfrac{1}{h^r} \sum_{\nu=0}^{\infty} b_{2\nu+r}^r\, \mu\delta^{2\nu+r}, & (8.2) \end{cases}$$

where $a_{2\nu+r}^r$ and $b_{2\nu+r}^r$ are numerical constants.†

† See, for example, Bickley, 1948, where general expressions for $a_{2\nu+r}^r$ and $b_{2\nu+r}^r$ are given for any value of r and for $0 \leqslant \nu \leqslant 5$.

Table (7.15)

$w(x)$	$U_r(x_k) \quad (r \geqslant 1)$

$$\cosh(\alpha x): \quad \tfrac{1}{2}\{e^{\alpha x_k}\chi_r(u)+e^{-\alpha x_k}\chi_r(-u)\} = (-)^r \begin{cases} \dfrac{\sinh\left\{\alpha\left(x_k+\dfrac{r-1}{2}h\right)\right\}}{\{2\sinh(u/2)\}^{r+1}} & (r\text{ even}) \\[4ex] \dfrac{\cosh\left\{\alpha\left(x_k+\dfrac{r-1}{2}h\right)\right\}}{\{2\sinh(u/2)\}^{r+1}} & (r\text{ odd}) \end{cases} + \tfrac{1}{2}\{e^{\alpha x_k}g_r(u)+e^{-\alpha x_k}g_r(-u)\}$$

$$\sinh(\alpha x): \quad \tfrac{1}{2}\{e^{\alpha x_k}\chi_r(u)-e^{-\alpha x_k}\chi_r(-u)\} = (-)^r \begin{cases} \dfrac{\cosh\left\{\alpha\left(x_k+\dfrac{r-1}{2}h\right)\right\}}{\{2\sinh(u/2)\}^{r+1}} & (r\text{ even}) \\[4ex] \dfrac{\sinh\left\{\alpha\left(x_k+\dfrac{r-1}{2}h\right)\right\}}{\{2\sinh(u/2)\}^{r+1}} & (r\text{ odd}) \end{cases} + \tfrac{1}{2}\{e^{\alpha x_k}g_r(u)-e^{-\alpha x_k}g_r(-u)\}$$

$$\cos(\beta x): \quad \chi_r^{\mathrm{I}}(iv)\cos(\beta x_k)+i\chi_r^{\mathrm{II}}(iv)\sin(\beta x_k) = \begin{cases} \dfrac{(-)^{r/2}\sin\left\{\beta\left(x_k+\dfrac{r-1}{2}h\right)\right\}}{\{2\sin(v/2)\}^{r+1}} & (r\text{ even}) \\[4ex] \dfrac{(-)^{(r-1)/2}\cos\left\{\beta\left(x_k+\dfrac{r-1}{2}h\right)\right\}}{\{2\sin(v/2)\}^{r+1}} & (r\text{ odd}) \end{cases} + g_r^{\mathrm{I}}(iv)\cos(\beta x_k)+ig_r^{\mathrm{II}}(iv)\sin(\beta x_k)$$

$$\sin(\beta x): \quad \chi_r^{\mathrm{I}}(iv)\sin(\beta x_k)-i\chi_r^{\mathrm{II}}(iv)\cos(\beta x_k) = \begin{cases} \dfrac{(-)^{1+r/2}\cos\left\{\beta\left(x_k+\dfrac{r-1}{2}h\right)\right\}}{\{2\sin(v/2)\}^{r+1}} & (r\text{ even}) \\[4ex] \dfrac{(-)^{(r-1)/2}\sin\left\{\beta\left(x_k+\dfrac{r-1}{2}h\right)\right\}}{\{2\sin(v/2)\}^{r+1}} & (r\text{ odd}) \end{cases} + g_r^{\mathrm{I}}(iv)\sin(\beta x_k)-ig_r^{\mathrm{II}}(iv)\cos(\beta x_k)$$

Table (7.16)

$w(x)$	$V_r(x_k) \quad (r \geqslant 1)$

$$\cosh(\alpha x) \qquad \frac{(-1)^{r+1}}{2}\{e^{\alpha x_k}\chi_r(-u) + e^{-\alpha x_k}\chi_r(u)\} = (-)^r \left\{ \begin{array}{l} \dfrac{\sinh\left\{\alpha\left(x_k - \dfrac{r-1}{2}h\right)\right\}}{\{2\sinh(u/2)\}^{r+1}} \quad (r \text{ even}) \\[2em] \dfrac{\cosh\left\{\alpha\left(x_k - \dfrac{r-1}{2}h\right)\right\}}{\{2\sinh(u/2)\}^{r+1}} \quad (r \text{ odd}) \end{array} \right\} + \frac{(-1)^{r+1}}{2}\{e^{\alpha x_k}g_r(-u) + e^{-\alpha x_k}g_r(u)\}$$

$$\sinh(\alpha x) \qquad \frac{(-1)^{r+1}}{2}\{e^{\alpha x_k}\chi_r(-u) - e^{-\alpha x_k}\chi_r(u)\} = (-)^r \left\{ \begin{array}{l} \dfrac{\cosh\left\{\alpha\left(x_k - \dfrac{r-1}{2}h\right)\right\}}{\{2\sinh(u/2)\}^{r+1}} \quad (r \text{ even}) \\[2em] \dfrac{\sinh\left\{\alpha\left(x_k - \dfrac{r-1}{2}h\right)\right\}}{\{2\sinh(u/2)\}^{r+1}} \quad (r \text{ odd}) \end{array} \right\} + \frac{(-1)^{r+1}}{2}\{e^{\alpha x_k}g_r(-u) - e^{-\alpha x_k}g_r(u)\}$$

$$\cos(\beta x) \qquad (-)^{r+1}\{\chi_r^{\mathrm{I}}(iv)\cos(\beta x_k) - i\chi_r^{\mathrm{II}}(iv)\sin(\beta x_k)\} = \left\{ \begin{array}{l} \dfrac{(-)^{r/2}\sin\left\{\beta\left(x_k - \dfrac{r-1}{2}h\right)\right\}}{\{2\sin(v/2)\}^{r+1}} \quad (r \text{ even}) \\[2em] \dfrac{(-)^{(r-1)/2}\cos\left\{\beta\left(x_k - \dfrac{r-1}{2}h\right)\right\}}{\{2\sin(v/2)\}^{r+1}} \quad (r \text{ odd}) \end{array} \right\} + (-1)^{r+1}\{g_r^{\mathrm{I}}(iv)\cos(\beta x_k) - ig_r^{\mathrm{II}}(iv)\sin(\beta x_k)\}$$

$$\sin(\beta x) \qquad (-)^{r+1}\{\chi_r^{\mathrm{I}}(iv)\sin(\beta x_k) + i\chi_r^{\mathrm{II}}(iv)\cos(\beta x_k)\} = \left\{ \begin{array}{l} \dfrac{(-)^{1+r/2}\cos\left\{\beta\left(x_k - \dfrac{r-1}{2}h\right)\right\}}{\{2\sin(v/2)\}^{r+1}} \quad (r \text{ even}) \\[2em] \dfrac{(-)^{(r-1)/2}\sin\left\{\beta\left(x_k - \dfrac{r-1}{2}h\right)\right\}}{\{2\sin(v/2)\}^{r+1}} \quad (r \text{ odd}) \end{array} \right\} + (-1)^{r+1}\{g_r^{\mathrm{I}}(iv)\sin(\beta x_k) + ig_r^{\mathrm{II}}(iv)\cos(\beta x_k)\}$$

With the help of these formulae the correction terms in a trapezoidal formula may be appropriately expressed in terms of central differences and their means by replacing the derivatives of even order, by using (8.1), and those of odd order, by using (8.2).

Thus we may derive, corresponding to (3.14) and (4.7) of Chapter I, the *trapezoidal formulae* with the weighting functions e^{px}, $\cosh(\alpha x)$, $\sinh(\alpha x)$, $\cos(\beta x)$ and $\sin(\beta x)$ in the form

$$h\left\{\tfrac{1}{2}w(x_0)f(x_0) + \sum_{r=1}^{n-1} w(x_r)f(x_r) + \tfrac{1}{2}w(x_n)f(x_n)\right\} - \int_{x_0}^{x_n} w(x)f(x)\,\mathrm{d}x$$

$$= h\sum_{r=0}^{\infty}\left[U_{2r}\{V_{2r}(x_n)\,\delta^{2r}f(x_n) - V_{2r}(x_0)\,\delta^{2r}f(x_0)\} \right.$$

$$\left. + U_{2r+1}\{V_{2r+1}(x_n)\,\mu\delta^{2r+1}f(x_n) - V_{2r+1}(x_0)\,\mu\delta^{2r+1}f(x_0)\}\right], \quad (8.3)$$

where, using the notations

$$\xi_0(q) = D_0(0,q) + \tfrac{1}{2} = \sum_{j=0}^{\infty}\frac{B_{2j+2}}{(2j+2)!}\,q^{2j+1}, \tag{8.4}$$

$$\xi_{2r}(q) = \sum_{\nu=1}^{r}\frac{a_{2r}^{2\nu}D_{2\nu}(0,q)}{(2\nu)!} = \sum_{j=0}^{\infty} c_{2j+1}^{2r}\,q^{2j+1} \quad (r\geqslant 1), \tag{8.5}$$

where

$$c_{2j+1}^{2r} = \frac{1}{2(2j+1)!}\sum_{\nu=1}^{r}\frac{a_{2r}^{2\nu}B_{2j+2\nu+2}}{(2\nu)!\,(j+\nu+1)}, \tag{8.6}$$

and

$$\xi_{2r+1}(q) = \sum_{\nu=0}^{r}\frac{b_{2r+1}^{2\nu+1}D_{2\nu+1}(0,q)}{(2\nu+1)!} = \sum_{j=0}^{\infty} c_{2j}^{2r+1}\,q^{2j}, \tag{8.7}$$

where

$$c_{2j}^{2r+1} = \frac{1}{2(2j)!}\sum_{\nu=0}^{r}\frac{b_{2r+1}^{2\nu+1}B_{2j+2\nu+2}}{(2\nu+1)!\,(j+\nu+1)}, \tag{8.8}$$

the coefficients U_r and $V_r(x_k)$ $(k=0,n)$, corresponding to each weighting function, are given below in table (8.16).

A mid-ordinate formula may similarly be transformed by replacing the derivatives of even order, by using (8.2), and those of odd order, by using (8.1).

Thus we may obtain, corresponding to (3.12) and (4.4) of Chapter I, the *mid-ordinate formulae* with the weighting functions e^{px}, $\cosh(\alpha x)$, $\sinh(\alpha x)$, $\cos(\beta x)$ and $\sin(\beta x)$ in the form

$$h\sum_{r=0}^{n-1} w(x_r+h/2)f(x_r+h/2) - \int_{x_0}^{x_n} w(x)f(x)\,\mathrm{d}x$$

$$= h\sum_{r=0}^{\infty}[W_{2r}\{V_{2r}(x_n)\,\mu\delta^{2r}f(x_n) - V_{2r}(x_0)\,\mu\delta^{2r}f(x_0)\}$$

$$+ W_{2r+1}\{V_{2r+1}(x_n)\,\delta^{2r+1}f(x_n) - V_{2r+1}(x_0)\,\delta^{2r+1}f(x_0)\}], \quad (8.9)$$

where, using the notations

$$B_\nu(\tfrac{1}{2}) = (2^{1-\nu} - 1) B_\nu, \tag{8.10}$$

$$\eta_0(q) = e^{q/2} D_0(\tfrac{1}{2}, q) = \sum_{j=0}^{\infty} \frac{B_{2j+2}(\tfrac{1}{2})}{(2j+2)!} q^{2j+1}, \tag{8.11}$$

$$\eta_{2r}(q) = \sum_{\nu=1}^{r} \frac{b_{2r}^{2\nu} e^{q/2} D_{2\nu}(\tfrac{1}{2}, q)}{(2\nu)!} = \sum_{j=0}^{\infty} d_{2j+1}^{2r} q^{2j+1} \quad (r \geqslant 1), \tag{8.12}$$

where

$$d_{2j+1}^{2r} = \frac{1}{2(2j+1)!} \sum_{\nu=1}^{r} \frac{b_{2r}^{2\nu} B_{2j+2\nu+2}(\tfrac{1}{2})}{(2\nu)! \, (j+\nu+1)}, \tag{8.13}$$

and

$$\eta_{2r+1}(q) = \sum_{\nu=0}^{r} \frac{a_{2r+1}^{2\nu+1} e^{q/2} D_{2\nu+1}(\tfrac{1}{2}, q)}{(2\nu+1)!} = \sum_{j=0}^{\infty} d_{2j}^{2r+1} q^{2j}, \tag{8.14}$$

where

$$d_{2j}^{2r+1} = \frac{1}{2(2j)!} \sum_{\nu=0}^{r} \frac{a_{2r+1}^{2\nu+1} B_{2j+2\nu+2}(\tfrac{1}{2})}{(2\nu+1)! \, (j+\nu+1)}, \tag{8.15}$$

the coefficients W_r and $V_r(x_k)$ $(k = 0, n)$, corresponding to each weighting function, are given below in table (8.16).

Table (8.16)

$$(ph = q, \alpha h = u, \beta h = v)$$

$w(x)$	$V_{2r}(x_k)$	$V_{2r+1}(x_k)$	U_r	W_r
e^{px}	e^{px_k}	e^{px_k}	$\xi_r(q)$	$\eta_r(q)$
$\cosh(\alpha x)$	$\sinh(\alpha x_k)$	$\cosh(\alpha x_k)$	$\xi_r(u)$	$\eta_r(u)$
$\sinh(\alpha x)$	$\cosh(\alpha x_k)$	$\sinh(\alpha x_k)$	$\xi_r(u)$	$\eta_r(u)$
$\cos(\beta x)$	$i \sin(\beta x_k)$	$\cos(\beta x_k)$	$\xi_r(iv)$	$\eta_r(iv)$
$\sin(\beta x)$	$-i \cos(\beta x_k)$	$\sin(\beta x_k)$	$\xi_r(iv)$	$\eta_r(iv)$

It may be noted here that generalizations, similar to (7.17), of (8.3) and (8.9) for the case of

$$\int_{x_0}^{x_n} w(x + \varepsilon) f(x) \, dx$$

are obtained by replacing, in each formula, $w(x)$ and $V_r(x_k)$ by $w(x+\varepsilon)$ and $V_r(x_k+\varepsilon)$ respectively.

9. *Numerical examples*

To illustrate the formulae of Sections 7 and 8, we have evaluated the two integrals

$$I(\alpha) = \int_0^\pi e^{\alpha x} \cos x \, dx$$

and

$$I(\beta) = \int_0^\pi \cos(\beta x) \, e^x \, dx$$

by using (i) the central difference formula (8.3), (ii) the formula (7.7) (or (7.12), as applicable) with forward differences at x_0 and backward differences at x_n, (iii) the Gregory formula and (iv) the Gauss formula.

The results obtained on neglecting all differences higher than the eighth are given below to eight significant figures.

(a) $\alpha = 8$, $\quad h = \pi/8$, $\quad u = \pi$, $\quad 10^3 I(\alpha) = -10120162$.

	Formula (7.7)	Gregory	Formula (8.3)	Gauss
Trapezoidal sum	-17477902	-17477902	-17477902	-17477902
δ^0	7325219	0	7325219	0
δ^1	133013	2583419	0	28710410
δ^2	-93291	1239771	32035	0
δ^3	-12459	753393	0	-82870748
δ^4	4182	513476	475	0
δ^5	1422	374706	0	170000990
δ^6	-201	286589	10	0
δ^7	-164	216276	0	260520370
δ^8	0	364873	0	0
Σ	-10120180	-11145399	-10120162	$+358883130$

(b) $\beta = 6$, $h = \pi/16$, $v = 3\pi/8$, $I(\beta) = 0.59839710$.

	Formula (7.12)	Gregory	Formula (8.3)	Gauss
Trapezoidal sum	0.67468915	0.67468915	0.67468915	0.67468915
δ^0	0	0	0	0
δ^1	-6864680	-26831411	-7684634	-2739658
δ^2	-687201	2814651	0	0
δ^3	-66252	11320136	55878	-2896587
δ^4	-9798	9312716	0	0
δ^5	-1058	3007176	-453	-1292425
δ^6	-191	-2516380	0	0
δ^7	-20	-4499152	4	-475500
δ^8	-4	-3125891	0	0
Σ	0.59839710	0.56950759	0.59839710	0.60064744

(c) $\beta = 12$, $h = \pi/16$, $v = 3\pi/4$, $I(\beta) = 0.15269443$.

	Formula (7.12)	Gregory	Formula (8.3)	Gauss
Trapezoidal sum	0.24880273	0.24880273	0.24880273	0.24880273
δ^0	0	0	0	0
δ^1	-8653979	-62908830	-9687656	5062228
δ^2	-866322	-43158772	0	0
δ^3	-78490	-29382060	77485	1856150
δ^4	-10754	-10016870	0	0
δ^5	-1079	22301895	-664	353718
δ^6	-183	75074371	0	0
δ^7	-18	1.55343230	6	-466233
δ^8	-4	2.66111670	0	0
Σ	0.15269444	3.98244910	0.15269443	0.31686137

10. *The forward, backward and central difference formulae for*

$$\int_{x_0}^{x_n} (x-x_0)^{\alpha_0}(x_n-x)^{\alpha_n} f(x)\, \mathrm{d}x$$

If we employ the procedure of Sections 7 and 8 to replace the derivatives in formula (8.3) of Chapter I by their forward, backward and central difference equivalents, we obtain the corresponding *trapezoidal formulae* with the weighting functions $w_0(x) = (x-x_0)^{\alpha_0}$ at x_0 and $w_n(x) = (x_n-x)^{\alpha_n}$ at x_n and with the correction terms expressed in terms of the forward and/or the backward and/or the central differences.

Using the notations

$$F(x) = (x-x_0)^{\alpha_0}(x_n-x)^{\alpha_n} f(x), \quad f_0(x) = (x_n-x)^{\alpha_n} f(x)$$

and

$$f_n(x) = (x-x_0)^{\alpha_0} f(x), \tag{10.1}$$

these formulae may be expressed in the form

$$
h\sum_{r=1}^{n-1} F(x_r) - \int_{x_0}^{x_n} F(x)\,\mathrm{d}x =
\begin{cases}
h^{1+\alpha_n} \sum_{r=0}^{\infty} V_r(\alpha_n)\, \Delta^r f_n(x_n) \\
\qquad\text{or} \\
h^{1+\alpha_n} \sum_{r=0}^{\infty} (-)^r U_r(\alpha_n)\, \nabla^r f_n(x_n) \\
\qquad\text{or} \\
h^{1+\alpha_n} \sum_{r=0}^{\infty} \{W_{2r}(\alpha_n)\, \delta^{2r} f_n(x_n) \\
\qquad\qquad - W_{2r+1}(\alpha_n)\, \mu\delta^{2r+1} f_n(x_n)\}
\end{cases}
$$

$$
\qquad +
\begin{cases}
h^{1+\alpha_0} \sum_{r=0}^{\infty} U_r(\alpha_0)\, \Delta^r f_0(x_0) \\
\qquad\text{or} \\
h^{1+\alpha_0} \sum_{r=0}^{\infty} (-)^r V_r(\alpha_0)\, \nabla^r f_0(x_0) \\
\qquad\text{or} \\
h^{1+\alpha_0} \sum_{r=0}^{\infty} \{W_{2r}(\alpha_0)\, \delta^{2r} f_0(x_0) \\
\qquad\qquad + W_{2r+1}(\alpha_0)\, \mu\delta^{2r+1} f_0(x_0)\}
\end{cases}, \tag{10.2}
$$

where, for $k = 0$ and $k = n$,

$$U_0(\alpha_k) = V_0(\alpha_k) = W_0(\alpha_k) = \zeta(-\alpha_k), \tag{10.3}$$

$$U_r(\alpha_k) = \frac{1}{r!} \sum_{\nu=1}^{r} S_r^\nu\, \zeta\{-(\nu+\alpha_k)\} \quad (r \geqslant 1), \tag{10.4}$$

$$V_r(\alpha_k) = \frac{1}{r!} \sum_{\nu=1}^{r} (-)^\nu S_r^\nu\, \zeta\{-(\nu+\alpha_k)\} \quad (r \geqslant 1), \tag{10.5}$$

$$W_{2r}(\alpha_k) = \sum_{\nu=1}^{r} \zeta\{-(2\nu+\alpha_k)\}\, a_{2r}^{2\nu}/(2\nu)! \quad (r \geqslant 1) \tag{10.6}$$

and

$$W_{2r+1}(\alpha_k) = \sum_{\nu=0}^{r} \zeta\{-(2\nu+1+\alpha_k)\}\, b_{2r+1}^{2\nu+1}/(2\nu+1)! \quad (r \geqslant 0). \tag{10.7}$$

Tables of $U_r(\alpha)$, $V_r(\alpha)$ and $W_r(\alpha)$, for $0 \leqslant r \leqslant 8$, are given in Appendix C for $\alpha = \pm\frac{1}{4}$, $\pm\frac{1}{3}$, $\pm\frac{1}{2}$, $\pm\frac{2}{3}$ and $\pm\frac{3}{4}$.

In illustration, we have applied (10.2) with (i) forward differences at x_0 and backward differences at x_n and (ii) central differences at both x_0 and x_n to evaluate the elliptic integral

$$\int_{-1}^{1} \sqrt{\{(1-x^2)(2-x)\}}\, dx = 2.2033457\ldots,$$

which has been discussed as a 'difficult' integral (Scarborough, 1958, p. 162). The results obtained for $h = 0.2$ and on neglecting all differences of orders higher than the eighth are given below.

	(i)	(ii)
Trapezoidal sum	2.1317965	2.1317965
δ^0	718413	718413
δ^1	-3397	-3023
δ^2	447	125
δ^3	13	-23
δ^4	14	-2
δ^5	-1	3
δ^6	3	0
δ^7	-3	-1
δ^8	7	0
$\sum$	2.2033462	2.2033457

It may be observed that in the case (i) the best result is obtained by neglecting all differences of orders higher than the fourth.

A METHOD BASED ON THE PROPERTIES OF THE FUNCTION $\Phi(z,\mu,\nu)$

In the following we briefly describe the method for expressing the correction coefficients for the weighting function $w_0(z) = (z - x_0)^{\alpha_0}\,\mathrm{e}^{pz}$ in terms of the higher transcendental function $\Phi(z,\mu,\nu)$. For $\alpha_0 = 0$ the treatment provides us with an alternative method for obtaining the coefficients $L_r(s,q)$ in the formula (3.7) of Chapter I.

The properties of the Φ-function

The function $\Phi(z,\mu,\nu)$ (see *Higher transcendental functions*, vol. I for (A1–5)), which may be formally defined as

$$\Phi(z,\mu,\nu) = \sum_{j=0}^{\infty} (\nu+j)^{-\mu}\, z^j \tag{A1}$$

(strictly valid for (i) $|z| < 1$ and either $\nu \neq 0, -1, -2, \ldots$ and any μ, or any ν and $\mu \leqslant 0$; (ii) either $|z| = 1$, $z \neq 1$, $\nu \neq 0, -1, -2, \ldots$, $\mu \geqslant 1$, or $z = 1$, $\nu \neq 0$, $-1, -2, \ldots$, $\mu > 1$), has the integral formula

$$\Phi(z,\mu,\nu) = \frac{1}{\Gamma(\mu)} \int_0^\infty \frac{t^{\mu-1}\,\mathrm{e}^{-\nu t}}{1 - z\,\mathrm{e}^{-t}}\, \mathrm{d}t \tag{A2}$$

($\mathrm{Re}\,\{\nu\} > 0$, and either $z \neq a$, where $a \geqslant 1$ is purely real, and $\mathrm{Re}\,\{\mu\} > 0$, or $z = 1$, and $\mathrm{Re}\,\{\mu\} > 1$), as well as the contour integral (Hankel type) representation

$$\Phi(z,\mu,\nu) = -\frac{\Gamma(1-\mu)}{2\pi\mathrm{i}} \int_\infty^{(0+)} \frac{(-t)^{\mu-1}\,\mathrm{e}^{-\nu t}}{1 - z\,\mathrm{e}^{-t}}\, \mathrm{d}t \tag{A3}$$

($\mathrm{Re}\,\{\nu\} > 0$, $z \neq a$, where $a \geqslant 1$ is purely real, and $\mu \neq 1, 2, 3, \ldots$).

We also have Lerch's transformation formula

$$\Phi(z,\mu,\nu) = \mathrm{i}z^{-\nu}(2\pi)^{\mu-1}\,\Gamma(1-\mu)\,\{\mathrm{e}^{-\mathrm{i}\pi\mu/2}\,\Phi(\mathrm{e}^{-2\pi\mathrm{i}\nu}, 1-\mu, \ln z/2\pi\mathrm{i})$$

$$- \mathrm{e}^{\mathrm{i}\pi(\mu/2+2\nu)}\,\Phi(\mathrm{e}^{2\pi\mathrm{i}\nu}, 1-\mu, 1-\ln z/2\pi\mathrm{i})\} \tag{A4}$$

$(0 < \nu \leqslant 1, \operatorname{Re}\{\mu\} < 0, \ z \neq a$, where $a \geqslant 1$ is purely real), and the Lindelöf–Wirtinger expansion

$$\Phi(z, \mu, \nu) = z^{-\nu}\,\Gamma(1-\mu)\,\{\ln(1/z)\}^{\mu-1} + z^{-\nu}\sum_{r=0}^{\infty}\zeta(\mu-r,\nu)\,(\ln z)^r/r! \qquad (A5)$$

$(|\ln z| < 2\pi, \ \mu \neq 1, 2, 3, \ldots, \ \nu \neq 0, -1, -2, \ldots)$.

It follows from (A1) that

$$\Phi(z, \mu, \nu) = z\,\Phi(z, \mu, 1+\nu) + \nu^{-\mu} \qquad (A6)$$

so that, taking $z = e^q$, $\mu = -(r+\alpha_0)$ and $\nu = s$, we may write Lerch's formula (A4) in the form

$$\Phi(e^q, -r-\alpha_0, s) = e^{-sq}(2\pi)^{-r-1-\alpha_0}\,\Gamma(r+1+\alpha_0)$$

$$\times\,[e^{i\pi(r+1+\alpha_0)/2}\{e^{-2\pi is}\,\Phi(e^{-2\pi is}, r+1+\alpha_0, 1-iq/2\pi)$$

$$+\,(-iq/2\pi)^{-r-1-\alpha_0}\} + e^{-i\pi(r+1+\alpha_0)/2}\,e^{2\pi is}$$

$$\times\,\Phi(e^{2\pi is}, r+1+\alpha_0, 1+iq/2\pi)]. \qquad (A7)$$

Another useful result is that when $r = 0, 1, 2, \ldots$,

$$\Phi(e^q, -r, s) = -C_r(s, q). \qquad (A8)$$

To prove this we note that if μ is an integer, the integrand of (A3) is a one-valued function of t, and we may apply Cauchy's theorem, and that (cf. Chapter II, eq. (1.1))

$$(-t)^{-r-1}\,\frac{e^{-st}}{1 - e^q\,e^{-t}} = (-)^r\sum_{m=0}^{\infty}\frac{t^{m-r-1}}{m!}\,(-)^m\,C_m(s, q),$$

so that, for $\mu = -r$ $(r = 0, 1, 2, \ldots)$, $z = e^q$ and $\nu = s$, the residue of the integrand of (A3) at $t = 0$ is $C_r(s, q)/r!$, and this proves (A8).

A different proof of (A8) is given later in Appendix B.

The weighting function $w_0(z) = (z - x_0)^{\alpha_0}\,e^{pz}$

Here $w_0(z) = (z - x_0)^{\alpha_0}\,e^{pz}$, so that

$$w_0(x_0 + iht/2\pi) = e^{px_0 + \alpha_0\pi i/2}\,e^{iqt/2\pi}(ht/2\pi)^{\alpha_0}, \qquad (A9)$$

and, from (2.9) of Chapter I,

$$A_{r0}(\xi, s) = (h/2\pi)^{\alpha_0}\,e^{px_0}\int_{\xi}^{\infty}\left\{\frac{(-)^r\,e^{-i\pi\alpha_0/2}\,e^{-iqt/2\pi}}{e^{-2\pi is}\,e^t - 1} - \frac{e^{i\pi\alpha_0/2}\,e^{iqt/2\pi}}{e^{2\pi is}\,e^t - 1}\right\}t^{r+\alpha_0}\,\mathrm{d}t. \qquad (A10)$$

Using (A2), (A10) with $\xi = 0$ may be written in the form

$$A_{r0}(0,s) = (h/2\pi)^{\alpha_0} e^{px_0} \Gamma(r+1+\alpha_0) e^{-i\pi(r-1)/2} \{ e^{-i\pi(r+1+\alpha_0)/2} e^{2\pi is}$$

$$\times \Phi(e^{2\pi is}, r+1+\alpha_0, 1+iq/2\pi) + e^{i\pi(r+1+\alpha_0)/2} e^{-2\pi is}$$

$$\times \Phi(e^{-2\pi is}, r+1+\alpha_0, 1-iq/2\pi) \}$$

$$= -(-2\pi i)^{r+1} h^{\alpha_0} e^{px_0} e^{sq}$$

$$\times \{ \Phi(e^q, -r-\alpha_0, s) - \Gamma(r+1+\alpha_0) e^{-sq}(-q)^{-r-1-\alpha_0} \}, \tag{A11}$$

from (A7).

Using (A11) in (2.12) of Chapter I, we have, finally, the desired expression for the correction coefficients $\phi_r^0(h,s)$, corresponding to the weighting function $w_0(z) = (z-x_0)^{\alpha_0} e^{pz}$, in the form

$$\phi_r^0(h,s) = -h^{r+1+\alpha_0} e^{px_0} e^{sq} \{ \Phi(e^q, -r-\alpha_0, s) - \Gamma(r+1+\alpha_0) e^{-sq}(-q)^{-r-1-\alpha_0} \}, \tag{A12}$$

as derived also by the operational method in Chapter III (Section 5).

The special case $\alpha_0 = 0$ corresponds to $w_0(z) = e^{pz}$ and then

$$\phi_r^0(h,s) = h^{r+1} e^{px_0} L_r(s,q),$$

in the notation of (3.7) of Chapter I, so that (A12) reduces to

$$L_r(s,q) = -e^{sq} \{ \Phi(e^q, -r, s) - r! \, e^{-sq}(-1/q)^{r+1} \}, \tag{A13}$$

and consequently, from (A8) and (3.1) of Chapter II, there follows (3.10) of Chapter I.

A PROBLEM OF UNSUMMABLE DIVERGENCE

The operational treatment of the right side of (4.1) of Chapter III gives rise to certain divergent series summation problems which are, in essence, the same as those encountered in the simpler problem of obtaining the expansion

$$\frac{t\,e^{st}}{e^t - 1} = \sum_{r=0}^{\infty} \frac{t^r}{r!} B_r(s), \tag{B1}$$

in ascending powers of t, by the following procedure:

$$\frac{t\,e^{st}}{e^t - 1} = -t \sum_{m=0}^{\infty} e^{(s+m)t} \tag{B2}$$

$$= -t \sum_{m=0}^{\infty} \sum_{r=0}^{\infty} \frac{(s+m)^r t^r}{r!}$$

$$= -t \sum_{r=0}^{\infty} \frac{t^r}{r!} \sum_{m=0}^{\infty} (s+m)^r, \tag{B3}$$

inverting the order of the summations.

The sum $\sum_{m=0}^{\infty}(s+m)^r$, however, is divergent for all $r \geqslant -1$. It is also difficult to assign any definite sum to these *truly* divergent series which are not summable in the ordinary sense.

If, as is 'natural' (see, e.g., Hardy, 1956, p. 346, Section 13.17), we assign to $\sum_{m=0}^{\infty}(s+m)^r$ the sum $\zeta(-r, s)$, which follows from the definition of the generalized Riemann zeta function, we obtain, from (B3), the expansion

$$\frac{t\,e^{st}}{e^t - 1} = -t \sum_{r=0}^{\infty} \frac{t^r}{r!} \zeta(-r, s)$$

$$= \sum_{r=1}^{\infty} \frac{t^r}{r!} B_r(s), \tag{B4}$$

which, although very nearly exact, is incorrect since it lacks the first term.

This difficulty is not eradicated by expanding in powers of e^{-t}, instead of e^t, at step (B2) above, since a follow through results only in the reproduction of the same erroneous result (B4).

The following procedure, which requires dealing with strictly convergent series only, may instead be pursued.

Assume $|z\,e^t| < 1$, then

$$-\frac{e^{st}}{z\,e^t - 1} = \sum_{m=0}^{\infty} z^m\, e^{(m+s)t}$$

$$= \sum_{m=0}^{\infty} z^m \sum_{r=0}^{\infty} \frac{(m+s)^r}{r!}\, t^r$$

$$= \sum_{r=0}^{\infty} \frac{t^r}{r!} \sum_{m=0}^{\infty} z^m(m+s)^r,$$

inverting the order of summations.

Hence (cf. (A1) of Appendix A),

$$-\frac{e^{st}}{z\,e^t - 1} = \sum_{r=0}^{\infty} \frac{t^r}{r!}\, \Phi(z, -r, s) \tag{B5}$$

$$= \sum_{r=0}^{\infty} \frac{t^r}{r!}\, z^{-s} \left\{ \Gamma(1+r) \left(\ln \frac{1}{z}\right)^{-(r+1)} + \zeta(-r,s) + O(\ln z) \right\}, \tag{B6}$$

using the Lindelöf–Wirtinger expansion (cf. (A5) of Appendix A).

But

$$\sum_{r=0}^{\infty} \Gamma(1-r) \frac{t^r}{r!} \left(\ln \frac{1}{z}\right)^{-(r+1)} = \frac{1}{\ln\,(1/z) - t},$$

so that (B6) reduces to

$$-\frac{e^{st}}{z\,e^t - 1} = z^{-s} \left\{ \frac{1}{\ln\,(1/z) - t} + \sum_{r=0}^{\infty} \frac{t^r}{r!}\, \zeta(-r,s) + O(\ln z) \right\},$$

where we may now let z tend to 1 and obtain, on multiplying both sides by $-t$, the expansion

$$\frac{t\,e^{st}}{e^t - 1} = 1 - t \sum_{r=0}^{\infty} \frac{t^r}{r!}\, \zeta(-r,s)$$

$$= \sum_{r=0}^{\infty} \frac{t^r}{r!}\, B_r(s), \tag{B7}$$

correctly.

It is interesting to notice that (B5) above, with $z = e^q$, yields (cf. (1.1) of Chapter II)

$$C_r(s,q) = -\Phi(e^q, -r, s),$$

as obtained also in Appendix A by the method of residues.

TABLES

Description of the tables and usage

For $r \geqslant 1$, values of $\lambda_r(q) = D_r(0, q)$ were computed from the asymptotic formula of Chapter II (Section 6), while for $r = 0$ use was made of the formula

$$\lambda_0(q) = D_0(0, q) + \tfrac{1}{2} = \frac{1}{2}\left(\frac{e^q + 1}{e^q - 1}\right) - \frac{1}{q}.$$

Satisfactory results were also obtained by an algorithm based on analytic continuation using the series expansion (4.13) of Chapter II for $z = 0$:

$$D_r(0, q + \varepsilon) = \sum_{\nu=0}^{\infty} \frac{\varepsilon^\nu}{\nu!} D_{r+\nu}(0, q), \tag{C1}$$

starting initially with the values for $q = 0$ given in terms of the Bernoulli numbers (see (4.11) of Chapter II).

For interpolation in Tables I and II the Taylor series (C1) above is recommended, the first four terms being sufficient for eight-figure accuracy.

Tables I and II give the eight-figure values of $\lambda_0(\pm q) = D_0(0, \pm q) + \tfrac{1}{2}$ and $\lambda_r(\pm q) = D_r(0, \pm q)$, $r = 1\,(1)\,11$ for $q = u$ and $q = iv$ and for u, $v = 0\,(0.02)\,1$. The relevant trapezoidal formulae are discussed in Sections 3, 4 and 6 of Chapter I. The D-functions are described in Chapter II.

Table III gives $\zeta\{-(r+\alpha)\}$, $W_r(\alpha)$, $U_r(\alpha)$ and $V_r(\alpha)$ for $r = 0\,(1)\,8$ and for $\alpha = \pm\tfrac{1}{4}$, $\pm\tfrac{1}{3}$, $\pm\tfrac{1}{2}$, $\pm\tfrac{2}{3}$ and $\pm\tfrac{3}{4}$. ζ is given to eight figures ando thers are given to eight decimals. The relevant trapezoidal and mid-ordinate formulae are discussed in Chapter I, Section 8 and Chapter IV, Section 10.

TABLE I

$$\lambda_0(u) = D_0(0, u) + \tfrac{1}{2}, \quad \lambda_r(u) = D_r(0, u) \quad (r \geq 1)$$
$$\lambda_r(-u) = (-)^{r+1} \lambda_r(u) \quad (\text{all } r)$$

u	$\lambda_0(u)$	$\lambda_1(u)$	$\lambda_2(u)$	$\lambda_3(u)$	$\lambda_4(u)$	$\lambda_5(u)$
0.00	0.0000000000	0.083333333	$-$0.00000000000	$-$0.0083333333	0.000000000000	0.0039682540
2	16666556	83331667	16666138	83325397	79359524	39674207
4	33332444	83326667	33329101	83301592	15868572	39649214
6	49997000	83318335	49985717	83261927	23794529	39607581
8	66659557	83306673	66632816	83206420	31710497	39549336
0.10	0.0083319448	0.083291683	$-$0.00083267230	$-$0.0083135094	0.00039613158	0.0039474522
2	99976008	83273368	99885801	83047979	47499205	39383193
4	11662857	83251730	11648537	82945111	55365339	39275417
6	13327648	83226775	13306280	82826533	63208280	39151270
8	14991906	83198507	14961494	82692295	71024762	39010843
0.20	0.016655566	0.083166931	$-$0.0016613868	$-$0.0082542454	0.00078811539	0.0038854238
2	18318561	83132053	18263089	82377071	86565386	38681568
4	19980826	83093881	19908847	82196216	94283103	38492957
6	21642295	83052420	21550835	81999965	10196151	38288541
8	23302901	83007680	23188744	81788398	10959748	38068468
0.30	0.024962580	0.082959668	$-$0.0024822269	$-$0.0081561605	0.0011718787	0.0037832896
2	26621266	82908394	26451107	81319679	12472961	37581993
4	28278894	82853867	28074956	81062721	13221966	37315940
6	29935399	82796098	29693516	80790837	13965499	37034926
8	31590717	82735097	31306491	80504140	14703264	36739152
0.40	0.033244782	0.082670876	$-$0.0032913584	$-$0.0080202747	0.0015434968	0.0036428828
2	34897530	82603447	34514503	79886783	16160322	36104174
4	36548898	82532822	36108959	79556378	16879041	35765420
6	38198822	82459015	37696663	79211668	17590846	35412806
8	39847238	82382040	39277331	78852792	18295463	35046578

TABLE I (cont.)

$$\lambda_0(u) = D_0(0, u) + \tfrac{1}{2}, \quad \lambda_r(u) = D_r(0, u) \quad (r \geqslant 1)$$
$$\lambda_r(-u) = (-)^{r+1} \lambda_r(u) \quad (\text{all } r)$$

u	$\lambda_0(u)$	$\lambda_1(u)$	$\lambda_2(u)$	$\lambda_3(u)$	$\lambda_4(u)$	$\lambda_5(u)$
0.50	0.041494083	0.082301911	−0.0040850681	−0.0078479899	0.0018992620	0.0034666994
2	43139293	82218643	42416435	78093139	19682055	34274319
4	44782807	82132250	43974316	77692670	20363508	33868827
6	46424563	82042751	45524051	77278654	21036725	33450800
8	48064497	81950160	47065373	76851258	21701458	33020526
0.60	0.049702548	0.081854495	−0.0048598014	−0.0076410654	0.0022357466	0.0032578302
2	51338656	81755774	50121712	75957019	23004512	32124432
4	52972759	81654014	51636209	75490534	23642368	31659226
6	54604797	81549235	53141249	75011387	24270808	31183001
8	56234709	81441456	54636581	74519766	24889616	30696079
0.70	0.057862435	0.081330696	−0.0056121958	−0.0074015868	0.0025498582	0.0030198790
2	59487917	81216975	57597135	73499890	26097501	29691466
4	61111094	81100314	59061874	72972036	26686176	29174447
6	62731910	80980734	60515939	72432512	27264417	28648077
8	64350304	80858258	61959098	71881530	27832039	28112704
0.80	0.065966221	0.080732905	−0.0063391125	−0.0071319303	0.0028388867	0.0027568680
2	67579602	80604701	64811797	70746048	28934731	27016361
4	69190390	80473666	66220895	70161988	29469469	26456106
6	70798530	80339825	67618206	69567345	29992925	25888277
8	72403964	80203201	69003520	68962347	30504952	25313240
0.90	0.074006639	0.080063819	−0.0070376632	−0.0068347224	0.0031005409	0.0024731362
2	75606499	79921703	71737343	67722209	31494163	24143012
4	77203489	79776878	73085456	67087536	31971089	23548560
6	78797556	79629370	74420781	66443445	32436067	22948380
8	80388646	79479203	75743133	65790174	32888988	22342843
1.00	0.081976707	0.079326406	−0.0077052329	−0.0065127966	0.0033329748	0.0021732324

TABLE I (cont.)

$$\lambda_0(u) = D_0(0, u) + \tfrac{1}{2}, \quad \lambda_r(u) = D_r(0, u) \quad (r \geqslant 1)$$
$$\lambda_r(-u) = (-)^{r+1} \lambda_r(u) \quad (\text{all } r)$$

u	$\lambda_6(u)$	$\lambda_7(u)$	$\lambda_8(u)$	$\lambda_9(u)$	$\lambda_{10}(u)$	$\lambda_{11}(u)$
0.00	-0.000000000000	-0.0041666667	0.00000000000	0.0075757576	-0.00000000000	-0.021092796
2	83323233	41651517	15148703	75715396	42174482	21076132
4	16658588	41606083	30280538	75588922	84282333	21026177
6	24972741	41530417	45378665	75378355	12625706	20943035
8	33268744	41424602	60426296	75084027	16803247	20826885
0.10	-0.00041540580	-0.0041288756	0.00075406723	0.0074706402	-0.0020954276	-0.020677972
2	49782255	41123031	90303343	74246076	25072272	20496613
4	57987812	40927610	10509969	73703774	29150784	20283193
6	66151334	40702710	11977944	73080350	33183442	20038162
8	74266948	40448581	13432648	72376784	37163976	19762041
0.20	-0.00082328836	-0.0040165503	0.0014872489	0.0071594180	-0.0041086224	-0.019455410
2	90331240	39853791	16295897	70733764	44944149	19118917
4	98268466	39513787	17701330	69796882	48731847	18753269
6	10613489	39145867	19087272	68784997	52443564	18359232
8	11392498	38750436	20452240	67699684	56073702	17937632
0.30	-0.0012163326	-0.0038327927	0.0021794781	0.0066542629	-0.0059616837	-0.017489347
2	12925438	37878804	23113479	65315625	63067724	17015309
4	13678304	37403556	24406952	64020569	66421310	16516503
6	14421409	36902703	25673861	62659454	69672742	15993956
8	15154245	36376788	26912904	61234371	72817381	15448746
0.40	-0.0015876317	-0.0035826381	0.0028122824	0.0059747500	-0.0075850804	-0.014881988
2	16587141	35252077	29302407	58201108	78768817	14294840
4	17286245	34654495	30450487	56597543	81567460	13688494
6	17973170	34034275	31565944	54939230	84243016	13064175
8	18647470	33392081	32647708	53228666	86792016	12423138

TABLE I (cont.)

$$\lambda_0(u) = D_0(0, u) + \tfrac{1}{2}, \quad \lambda_r(u) = D_r(0, u) \quad (r \geqslant 1)$$
$$\lambda_r(-u) = (-)^{r+1} \lambda_r(u) \quad (\text{all } r)$$

u	$\lambda_6(u)$	$\lambda_7(u)$	$\lambda_8(u)$	$\lambda_9(u)$	$\lambda_{10}(u)$	$\lambda_{11}(u)$
0.50	−0.0019308711	−0.0032728598	0.0033694759	0.0051468415	−0.0089211243	−0.011766666
2	19956476	32044529	34706131	49661102	91497740	11096063
4	20590360	31340597	35680908	47809408	93648814	10412654
6	21209973	30617542	36618230	45916068	95662037	97177808
8	21814939	29876123	37517291	43983860	97535252	90127981
0.60	−0.0022404900	−0.0029117111	0.0038377344	0.0042015604	−0.0099266574	−0.0082990706
2	22979512	28341294	39197694	40014154	10085439	75779693
4	23538445	27549472	39977708	37982394	10229736	68508686
6	24081389	26742458	40716807	35923232	10359443	61191424
8	24608048	25921076	41414474	33839595	10474480	53841613
0.70	−0.0025118142	−0.0025086159	0.0042070247	0.0031734422	−0.010574797	−0.0046472893
2	25611410	24238548	42683727	29610659	10660368	39098802
4	26087605	23379094	43254570	27471257	10731197	31732751
6	26546500	22508651	43782492	25319161	10787313	24387987
8	26987884	21628082	44267271	23157309	10828772	17077569
0.80	−0.0027411562	−0.0020738249	0.0044708739	0.0020988624	−0.010855655	−0.00098143353
2	27817358	19840022	45106790	18816012	10868069	26108770
4	28205112	18934267	45461373	16642353	10866147	0.00045204887
6	28574684	18021856	45772496	14470499	10850043	11567745
8	28925948	17103657	46040224	12303269	10819940	18519198
0.90	−0.0029258797	−0.0016180536	0.0046264676	0.0010143443	−0.010776038	0.0025363503
2	29573142	15253357	46446029	79937583	10718564	32089686
4	29868910	14322981	46584512	58569054	10647765	38687162
6	30146046	13390261	46680408	37355227	10563908	45145756
8	30404511	12456046	46734053	16321935	10467281	51455723
1.00	−0.0030644284	−0.0011521178	0.0046745833	−0.000045055878	−0.010358191	0.0057607761

TABLE II

$$\lambda_0(iv) = D_0(0, iv) + \tfrac{1}{2}, \quad \lambda_r(iv) = D_r(0, iv) \quad (r \geqslant 1)$$
$$\lambda_r(-iv) = (-)^{r+1} \lambda_r(iv) \quad (\text{all } r)$$

v	$\lambda_0(iv)/i$	$\lambda_1(iv)$	$\lambda_2(iv)/i$	$\lambda_3(iv)$	$\lambda_4(iv)/i$	$\lambda_5(iv)$
0.00	0.0000000000	0.083333333	-0.00000000000	-0.0083333333	0.000000000000	0.0039682540
2	16666778	83335000	16667196	83341270	79370635	39690874
4	33334222	83340000	33337566	83365084	15877461	39715881
6	50003000	83348335	50014288	83404784	23824529	39757581
8	66673779	83360007	66700540	83460389	31781608	39816002
0.10	0.0083347226	0.083375017	-0.00083399506	-0.0083531920	0.00039752047	0.0039891189
2	10002401	83393368	10011437	83619408	47739205	39983195
4	11670480	83415064	11684833	83722890	55746451	40092088
6	13339026	83440108	13360460	83842409	63777171	40217947
8	15008106	83468507	15038637	83978017	71834767	40360863
0.20	0.016677788	0.083500265	-0.0016719688	-0.0084129769	0.00079922660	0.0040520942
2	18348139	83535388	18403936	84297730	88044296	40698301
4	20019226	83573883	20091706	84481971	96203141	40893069
6	21691117	83615757	21783324	84682569	10440269	41105389
8	23363879	83661019	23479118	84899611	11264648	41335417
0.30	0.025037581	0.083709677	-0.0025179418	-0.0085133187	0.0012093805	0.0041583323
2	26712289	83761740	26884556	85383397	12928101	41849289
4	28388073	83817219	28594866	85650348	13767898	42133512
6	30065001	83876123	30310683	85934152	14613565	42436201
8	31743141	83938465	32032345	86234932	15465471	42757583
0.40	0.033422562	0.084004257	-0.0033760194	-0.0086552815	0.0016323904	0.0043097895
2	35103334	84073510	35494573	86887938	17189515	43457391
4	36785526	84146239	37235828	87240445	18062419	43836339
6	38469207	84222458	38984307	87610487	18943100	44235024
8	40154447	84302182	40740365	87998224	19831954	44653745

TABLE II (cont.)

$$\lambda_0(iv) = D_0(0, iv) + \tfrac{1}{2}, \quad \lambda_r(iv) = D_r(0, iv) \quad (r \geqslant 1)$$
$$\lambda_r(-iv) = (-)^{r+1} \lambda_r(iv) \quad (\text{all } r)$$

v	$\lambda_0(iv)/i$	$\lambda_1(iv)$	$\lambda_2(iv)/i$	$\lambda_3(iv)$	$\lambda_4(iv)/i$	$\lambda_5(iv)$
0.50	0.041841318	0.084385425	−0.0042504356	−0.0088403822	0.0020729385	0.0045092818
2	43529888	84472205	44276638	88827459	21635805	45552573
4	45220230	84562537	46057575	89269317	22551629	46033360
6	46912413	84656441	47847534	89729590	23477282	46535544
8	48606511	84753934	49646883	90208477	24413196	47059508
0.60	0.050302595	0.084855035	−0.0051455998	−0.0090706189	0.0025359810	0.0047605653
2	52000737	84959765	53275258	91222944	26317572	48174399
4	53701010	85068143	55105044	91758969	27286939	48766185
6	55403487	85180192	56945746	92314502	28268376	49381467
8	57108242	85295934	58797756	92889788	29262358	50020725
0.70	0.058815349	0.085415391	−0.0060661471	−0.0093485083	0.0030269369	0.0050684457
2	60524883	85538588	62537295	94100653	31289903	51373183
4	62236918	85665548	64425634	94736773	32324466	52087445
6	63951530	85796299	66326904	95393728	33373575	52827806
8	65668795	85930865	68241524	96071816	34437756	53594855
0.80	0.067388790	0.086069274	−0.0070169920	−0.0096771343	0.0035517551	0.0054389202
2	69111592	86211554	72112523	97492626	36613511	55211484
4	70837278	86357734	74069772	98235995	37726201	56062363
6	72565927	86507843	76042112	99001789	38856200	56942526
8	74297618	86661913	78029996	99790362	40004102	57852688
0.90	0.076032430	0.086819974	−0.0080033881	−0.010060208	0.0041170513	0.0058793593
2	77770444	86982059	82054235	10143731	42356056	59766012
4	79511740	87148202	84091533	10229645	43561369	60770748
6	81256399	87318437	86146256	10317990	44787106	61808634
8	83004505	87492799	88218894	10408808	46033941	62880535
1.00	0.084756139	0.087671325	−0.0090309946	−0.010502140	0.0047302561	0.0063987351

TABLE II (*cont.*)

$$\lambda_0(iv) = D_0(0, iv) + \tfrac{1}{2}, \quad \lambda_r(iv) = D_r(0, iv) \quad (r \geqslant 1)$$
$$\lambda_r(-iv) = (-)^{r+1} \lambda_r(iv) \quad (\text{all } r)$$

v	$\lambda_6(iv)/i$	$\lambda_7(iv)$	$\lambda_8(iv)/i$	$\lambda_9(iv)$	$\lambda_{10}(iv)/i$	$\lambda_{11}(iv)$
0.00	-0.000000000000	-0.0041666667	0.00000000000	0.0075757576	-0.00000000000	-0.021092796
2	83343435	41681820	15154328	75799767	42196705	21109466
4	16674749	41727295	30325536	75926407	84460111	21159510
6	25027286	41803144	45530534	76137696	12685706	21243036
8	33398037	41909451	60786280	76433969	16945469	21360220
0.10	-0.00041793105	-0.0042046335	0.00076109818	0.0076815694	-0.0021232055	-0.021511314
2	50218620	42213947	91518294	77283475	25552277	21696639
4	58680745	42412476	10702899	77838055	29913019	21916590
6	67185686	42642143	12265936	78480313	34321252	22171638
8	75739695	42903205	13842703	79211269	38784050	22462329
0.20	-0.00084349080	-0.0043195954	0.0015434986	0.0080032087	-0.0043308601	-0.022789287
2	93020211	43520720	17044594	80944073	47902229	23153212
4	10175953	43877866	18673366	81948686	52572403	23554890
6	11057354	44267795	20323170	83047531	57326759	23995186
8	11946886	44690946	21995908	84242370	62173116	24475054
0.30	-0.0012845217	-0.0045147797	0.0023693518	0.0085535122	-0.0067119488	-0.024995532
2	13753026	45638865	25417979	86927871	72174111	25557753
4	14671003	46164708	27171314	88422865	77345454	26162944
6	15599850	46725924	28955591	90022525	82642241	26812428
8	16540280	47323152	30772930	91729451	88073473	27507631
0.40	-0.0017493021	-0.0047957076	0.0032625503	0.0093546422	-0.0093648444	-0.028250085
2	18458813	48628422	34515540	95476411	99376766	29041431
4	19438412	49337963	36445334	97522581	10526839	29883426
6	20432591	50086516	38417241	99688304	11133364	30777946
8	21442139	50874949	40433687	10197716	11758321	31726994

TABLE II (cont.)

$$\lambda_0(iv) = D_0(0, iv) + \tfrac{1}{2}, \quad \lambda_r(iv) = D_r(0, iv) \quad (r \geqslant 1)$$
$$\lambda_r(-iv) = (-)^{r+1}\lambda_r(iv) \quad (\text{all } r)$$

v	$\lambda_6(iv)/i$	$\lambda_7(iv)$	$\lambda_8(iv)/i$	$\lambda_9(iv)$	$\lambda_{10}(iv)/i$	$\lambda_{11}(iv)$
0.50	−0.0022467861	−0.0051704177	0.0042497173	0.010439293	−0.012402821	−0.032732701
2	23510584	52575167	44610277	10693966	13068021	33797339
4	24571153	53488937	46775661	10962160	13755124	34923322
6	25650434	54446560	48996073	11244326	14465380	36113215
8	26749315	55449166	51274354	11540939	15200097	37369745
0.60	−0.0027868709	−0.0056497940	0.0053613445	0.011852502	−0.015960634	−0.038695802
2	29009549	57594129	56016387	12179546	16748413	40094454
4	30172799	58739042	58486332	12522630	17564917	41568954
6	31359445	59934051	61026547	12882344	18411699	43122750
8	32570504	61180595	63640420	13259310	19290380	44759493
0.70	−0.0033807022	−0.0062480182	0.0066331465	0.013654183	−0.020202658	−0.046483054
2	35070076	63834392	69103333	14067652	21150309	48297531
4	36360773	65244880	71959814	14500444	22135194	50207263
6	37680257	66713375	74904849	14953321	23159265	52216845
8	39029707	68241691	77942535	15427089	24224566	54331141
0.80	−0.0040410336	−0.0069831723	0.0081077134	0.015922593	−0.025333243	−0.056555300
2	41823400	71485452	84313081	16440723	26487547	58894772
4	43270193	73204952	87654994	16982415	27689841	61355326
6	44752051	74992390	91107683	17548653	28942607	63943069
8	46270357	76850032	94676161	18140473	30248454	66664463
0.90	−0.0047826536	−0.0078780244	0.0098365650	0.018758963	−0.031610122	−0.069526351
2	49422067	80785501	10218160	19405269	33030493	72535977
4	51058474	82868388	10612969	20080595	34512597	75701009
6	52737338	85031609	11021586	20786206	36059624	79029566
8	54460293	87277984	11444628	21523435	37674927	82530246
1.00	−0.0056229031	−0.0089610465	0.011882743	0.022293682	−0.039362042	−0.086212154

TABLE III

	r	$\zeta\{-(r+\alpha)\}$	$W_r(\alpha)$	$U_r(\alpha)$	$V_r(\alpha)$
$\alpha = -\frac{3}{4}$	0	-3.4412854	-3.44128539	-3.44128539	-3.44128539
	1	-0.32045126	-0.32045126	-0.32045126	$+0.32045126$
	2	-0.048908868	-0.02445443	$+0.13577120$	-0.18468007
	3	$+0.0057588940$	$+0.05436836$	-0.08140284	$+0.13031171$
	4	$+0.0066194766$	$+0.00231368$	$+0.05653234$	-0.10081385
	5	-0.0017912146	-0.01093659	-0.04259843	$+0.08225257$
	6	-0.0035570039	-0.00032262	$+0.03381019$	-0.06948222
	7	$+0.0014548168$	$+0.00235019$	-0.02781743	$+0.06015260$
	8	$+0.0040383564$	$+0.00005305$	$+0.02349810$	-0.05303556
$\alpha = -\frac{2}{3}$	0	-2.4475807	-2.44758074	-2.44758074	-2.44758074
	1	-0.27734305	-0.27734305	-0.27734305	$+0.27734305$
	2	-0.040061330	-0.02003066	$+0.11864086$	-0.15870219
	3	$+0.0069639515$	$+0.04738450$	-0.07125636	$+0.11131769$
	4	$+0.0059261453$	$+0.00191614$	$+0.04948025$	-0.08570929
	5	-0.0022842517	-0.00955397	-0.03725812	$+0.06965487$
	6	-0.0033092476	-0.00026831	$+0.02954465$	-0.05864574
	7	$+0.0019125487$	$+0.00205545$	-0.02428441	$+0.05062645$
	8	$+0.0038534280$	$+0.00004422$	$+0.02049387$	-0.04452508
$\alpha = -\frac{1}{2}$	0	-1.4603545	-1.46035451	-1.46035451	-1.46035451
	1	-0.20788622	-0.20788622	-0.20788622	$+0.20788622$
	2	-0.025485202	-0.01274260	$+0.09120051$	-0.11668571
	3	$+0.0085169288$	$+0.03606719$	-0.05513332	$+0.08061852$
	4	$+0.0044410113$	$+0.00124693$	$+0.03834665$	-0.06133800
	5	-0.0030916692	-0.00731018	-0.02887015	$+0.04936765$
	6	-0.0026714580	-0.00017614	$+0.02287261$	-0.04122853
	7	$+0.0027467679$	$+0.00157684$	-0.01877719	-0.03534380
	8	$+0.0032690396$	$+0.00002916$	$+0.01582462$	-0.03089587
$\alpha = -\frac{1}{3}$	0	-0.97336025	-0.97336025	-0.97336025	-0.97336025
	1	-0.15519690	-0.15519690	-0.15519690	$+0.15519690$
	2	-0.014373542	-0.00718677	$+0.07041168$	-0.08478522
	3	$+0.0091280360$	$+0.02738749$	-0.04302419	$+0.05739773$
	4	$+0.0029049761$	$+0.00071994$	$+0.03005038$	-0.04298405
	5	-0.0036454631	-0.00558394	-0.02266052	$+0.03415431$
	6	-0.0018766468	-0.00010263	$+0.01796000$	-0.02821918
	7	$+0.0034301980$	$+0.00120810$	-0.01474071	$+0.02397054$
	8	$+0.0024101679$	$+0.00001708$	$+0.01241569$	-0.02078729
$\alpha = -\frac{1}{4}$	0	-0.81327841	-0.81327841	-0.81327841	-0.81327841
	1	-0.13364277	-0.13364277	-0.13364277	$+0.13364277$
	2	-0.0099013776	-0.00495069	$+0.06187070$	-0.07177208
	3	$+0.0091471501$	$+0.02379832$	-0.03807238	$+0.04797376$
	4	$+0.0021445010$	$+0.00050191$	$+0.02667513$	-0.03557268
	5	-0.0038241109	-0.00486776	-0.02014564	$+0.02803937$
	6	-0.0014341298	-0.00007189	$+0.01597813$	-0.02301183
	7	$+0.0037006219$	$+0.00105488$	-0.01311774	$+0.01943518$
	8	$+0.0018860322$	$+0.00001199$	$+0.01104900$	-0.01676999

TABLE III (*cont.*)

	r	$\zeta\{-(r+\alpha)\}$	$W_r(\alpha)$	$U_r(\alpha)$	$V_r(\alpha)$
$\alpha = +\frac{1}{4}$	0	-0.32045126	-0.32045126	-0.32045126	-0.32045126
	1	-0.048908868	-0.04890887	-0.04890887	$+0.04890887$
	2	$+0.0057588940$	$+0.00287945$	$+0.02733388$	-0.02157499
	3	$+0.0066194766$	$+0.00925472$	-0.01807916	$+0.01232026$
	4	-0.0017912146	-0.00031459	$+0.01313721$	-0.00800749
	5	-0.0035570039	-0.00193575	-0.01013101	$+0.00563046$
	6	$+0.0014548168$	$+0.00004645$	$+0.00813913$	-0.00417486
	7	$+0.0040383564$	$+0.00042439$	-0.00673720	$+0.00321629$
	8	-0.0021925719	-0.00000788	$+0.00570514$	-0.00255044
$\alpha = +\frac{1}{3}$	0	-0.27734305	-0.27734305	-0.27734305	-0.27734305
	1	-0.040061330	-0.04006133	-0.04006133	$+0.04006133$
	2	$+0.0069639515$	$+0.00348198$	$+0.02351264$	-0.01654869
	3	$+0.0059261453$	$+0.00766458$	-0.01584806	$+0.00888411$
	4	-0.0022842517	-0.00038534	$+0.01163043$	-0.00543716
	5	-0.0033092476	-0.00160988	-0.00902268	$+0.00360009$
	6	$+0.0019125487$	$+0.00005721$	$+0.00727707$	-0.00251075
	7	$+0.0038534280$	$+0.00035372$	-0.00603988	$+0.00181542$
	8	-0.0029459105	-0.00000973	$+0.00512453$	-0.00134696
$\alpha = +\frac{1}{2}$	0	-0.20788622	-0.20788622	-0.20788622	-0.20788622
	1	-0.025485202	-0.02548520	-0.02548520	$+0.02548520$
	2	$+0.0085169288$	$+0.00425846$	$+0.01700107$	-0.00848414
	3	$+0.0044410113$	$+0.00498770$	-0.01201336	$+0.00349643$
	4	-0.0030916692	-0.00048369	$+0.00903582$	-0.00148627
	5	-0.0026714580	-0.00105681	-0.00711509	$+0.00053293$
	6	$+0.0027467679$	$+0.00007260$	$+0.00579536$	-0.00003538
	7	$+0.0032690396$	$+0.00023328$	-0.00484336	-0.00023964
	8	-0.0044160329	-0.00001242	$+0.00413002$	$+0.00039636$
$\alpha = +\frac{2}{3}$	0	-0.15519690	-0.15519690	-0.15519690	-0.15519690
	1	-0.014373542	-0.01437354	-0.01437354	$+0.01437354$
	2	$+0.0091280360$	$+0.00456402$	$+0.01175079$	-0.00262275
	3	$+0.0029049761$	$+0.00287975$	-0.00887104	-0.00025700
	4	-0.0036454631	-0.00053223	$+0.00689893$	$+0.00116465$
	5	-0.0018766468	-0.00061580	-0.00554262	-0.00145650
	6	$+0.0034301980$	$+0.00008079$	$+0.00457500$	$+0.00152124$
	7	$+0.0024101679$	$+0.00013660$	-0.00385947	-0.00149548
	8	-0.0057507340	-0.00001391	$+0.00331382$	$+0.00143360$
$\alpha = +\frac{3}{4}$	0	-0.13364277	-0.13364277	-0.13364277	-0.13364277
	1	-0.0099013776	-0.00990138	-0.00990138	$+0.00990138$
	2	$+0.0091471501$	$+0.00457358$	$+0.00952426$	-0.00037711
	3	$+0.0021445010$	$+0.00200765$	-0.00751662	-0.00163053
	4	-0.0038241109	-0.00054047	$+0.00597232$	$+0.00209389$
	5	-0.0014341298	-0.00043135	-0.00485938	-0.00212589
	6	$+0.0037006219$	$+0.00008251$	$+0.00404463$	$+0.00202473$
	7	$+0.0018860322$	$+0.00009593$	-0.00343214	-0.00188634
	8	-0.0063332711	-0.00001426	$+0.00295968$	$+0.00174443$

REFERENCES

1   Bickley, W. G. 'Difference and associated operators, with some applications', *J. Math. Phys.* **27** (1948).
2   Boole, G. *A treatise on the calculus of finite differences*, London (1880).
3   Dwight, H. B. *Mathematical tables of elementary and some higher mathematical functions*, New York (1958).
4   Hardy, G. H. *Divergent series*, Oxford (1956).
5   *Higher transcendental functions* (vol. I). Compiled by the Staff of the Bateman Manuscript Project (A. Erdelyi and others), New York (1953).
6   Hildebrand, F. B. *Introduction to numerical analysis*, New York (1956).
7   Jordan, C. *Calculus of finite differences*, New York (1947).
8   Mikusinski, J. *Operational calculus*, London (1959).
9   Milne-Thompson, L. M. *The calculus of finite differences*, London (1951).
10 Navot, I. 'An extension of the Euler–Maclaurin summation formula to functions with a branch singularity', *J. Math. Phys.* **40** (1961).
    —— 'A further extension of the Euler–Maclaurin summation formula', *J. Math. Phys.* **41** (1962).
    —— 'The Euler–Maclaurin functional for functions with a quasi-step discontinuity', *Maths. Comp.* **17** (1963).
11 Nörlund, N. E. *Vorlesungen über Differenzenrechnung*, Berlin (1924).
12 Scarborough, J. B. *Numerical mathematical analysis*, London (1958).
13 Steffensen, J. F. *Interpolation*, New York (1950).
14 Whittaker, E. T. and Watson, G. N. *Modern analysis*, Cambridge (1952).